Seminary on Geomatics, Civil and Environmental Engineering (2018 BGC)

E3S Web of Conferences Volume 63 (2018)

Olsztyn, Poland
21-23 June 2018

Editors:

G. Nykiel

ISBN: 978-1-5108-7477-0

Printed from e-media with permission by:

Curran Associates, Inc.
57 Morehouse Lane
Red Hook, NY 12571

Some format issues inherent in the e-media version may also appear in this print version.

No changes have been made to the content of these proceedings. There may be changes to pagination and minor adjustments for aesthetics.

Printed by Curran Associates, Inc. (2019)

For additional information, please contact EDP Sciences – Web of Conferences at the address below.

EDP Sciences – Web of Conferences
17, Avenue du Hoggar
Parc d'Activité de Courtabœuf
BP 112
F-91944 Les Ulis Cedex A
France

Phone: +33 (0) 1 69 18 75 75
Fax: +33 (0) 1 69 28 84 91

contact-edps@webofconferences.org

Additional copies of this publication are available from:

Curran Associates, Inc.
57 Morehouse Lane
Red Hook, NY 12571 USA
Phone: 845-758-0400
Fax: 845-758-2633
Email: curran@proceedings.com
Web: www.proceedings.com

TABLE OF CONTENTS

Statement of Peer review

In submitting conference proceedings to *Web of Conferences*, the editors of the proceedings certify to the Publisher that

1. They adhere to its **Policy on Publishing Integrity** in order to safeguard good scientific practice in publishing.
2. All articles have been subjected to peer review administered by the proceedings editors.
3. Reviews have been conducted by expert referees, who have been requested to provide unbiased and constructive comments aimed, whenever possible, at improving the work.
4. Proceedings editors have taken all reasonable steps to ensure the quality of the materials they publish and their decision to accept or reject a paper for publication has been based only on the merits of the work and the relevance to the journal.

Title, date and place of the conference

Seminary on Geomatics, Civil and Environmental Engineering (2018 BGC)
21-23 June 2018, University of Warmia and Mazury in Olsztyn, Poland

Proceedings editor(s):

Dr Grzegorz Nykiel Gdansk University of Technology, Poland

Date and editor's signature 30/10/2018

https://doi.org/10.1051/e3sconf/20186300001

A role of an Enterprise Identification Card in a building process of water tourist investments

Emilia Miszewska[1,*], *Adam* Bolt[2] , and *Magdalena* Apollo[1]

[1]Gdańsk University of Technology, Faculty of Civil and Environmental Engineering, Department of Construction Management and Earthquake Engineering, Narutowicza 11/12 Str., 80-233 Gdańsk, Poland
[2]Gdańsk University of Technology, Faculty of Civil and Environmental Engineering, Department of Geotechnics, Geology and Marine Civil Engineering, Narutowicza 11/12 Str., 80-233 Gdańsk, Poland

Abstract. An essential element of a building process for water tourist investments in Poland is a document called the Enterprise Identification Card. It is required for the investments, which are classified according to the Regulation [1] as the ones that could potentially impact the environment. It is assumed that KIP is there to protect the environment against the excessive human interference and it just provides the guidelines regarding the enterprise completion for an applicant. TOWS-SWOT analysis has been used in order to conduct an accurate evaluation of a role of the Enterprise Identification Card in a building process of water tourist investments. It is one of the modifications of the basic method of strategic analysis SWOT.

1 Introduction

An essential element of a building process for most water tourist investments in Poland is a necessity to obtain a decision about the environmental conditions (*an environmental decision* for short). In order to be granted the environmental decision, the Enterprise Identification Card (*KIP* for short) should be prepared in the correct way.

A decision about the environmental conditions is a document, which defines the environmental conditions for the completion of an enterprise in the most environment friendly way. In the documents an enterprise is described as a building plan or another environmental impact, which transforms or changes the way of an area is used, including ore extraction.

An enterprise information card is a document, which should include basic information about a planned building enterprise. It should characterize an enterprise in such a way that the information included allows to identify an investment, its scope and some basic parameters. KIP is required when a planned enterprise is numbered among the enterprises, which could potentially have a significant impact on the environment. It is a group of enterprises defined in §3 of The Council of Ministers Regulation on the enterprises, which could have a significant environmental impact (Journal of Laws 2010, No.213, item 1397 as amended) [1] and in Appendix II of the Council Directive No 85/337/EWG about an assessment of the impact of some enterprises on public and private environment [2].

KIP should also be obtained in the case when an investor applies for a decision about the environmental conditions for an enterprise which could have a significant impact on the environment. These are included in §2 of the above-mentioned regulation [1] and in the Appendix of I Council Directive No 85/337/EWG [2], however, it is only the case when an applicant does not include a report and requests removal of a scope of the report from the application [3].

The aim of the publication is to indicate a positive and a negative role of an Enterprise Identification Card and its strengths and weaknesses in a building process by means of TOWS-SWOT analysis for water tourist investments.

2 A role of an Enterprise Identification Card in an investment process

Figures and tables, as originals of good quality and well contrasted, are to be in their final form, ready for reproduction, pasted in the appropriate place in the text. Try to ensure that the size of the text in your figures is approximately the same size as the main text (10 point). Try to ensure that lines are no thinner than 0.25 point. An Enterprise Identification Card should be prepared for the investments, which could potentially have a significant impact on the environment. Water tourist investments such as a construction or reconstruction of canoe harbours and marinas, a construction and reconstruction of platforms and water tram stops are usually classified as this group.

In most procedures, an investor obliges a designer to obtain an environmental decision and to prepare KIP. As the research shows, waiting time for granting an environmental decision is [4] the main reason for the extension of an investment process. The problem is multifaceted and results from the incorrect preparation of KIPs, the wrong enterprise classification as the ones

* Corresponding author: emiurban@pg.edu.pl

which could have significant impact on the environment, too many cases for each official, lack of technical knowledge among the officials employed to issue a decision, or various law interpretations among the institutions involved in a process of issuing of an environmental decision. As a result, not only an construction process is extended, but there could also be a necessity to prepare a report about an impact of an enterprise on the environment and a refusal to issue an environmental decision, which is essential in order to obtain a construction permit or to report construction work for building investments in Poland. Problems with obtaining an environmental decision result in imposing large financial penalties on a designer by an Investor or/and impairment of a subsidy from external sources by an Investor and suspension or abandonment of an enterprise. The process of obtaining an environmental decision and a role of an Enterprise Identification Card is presented on Figure 1.

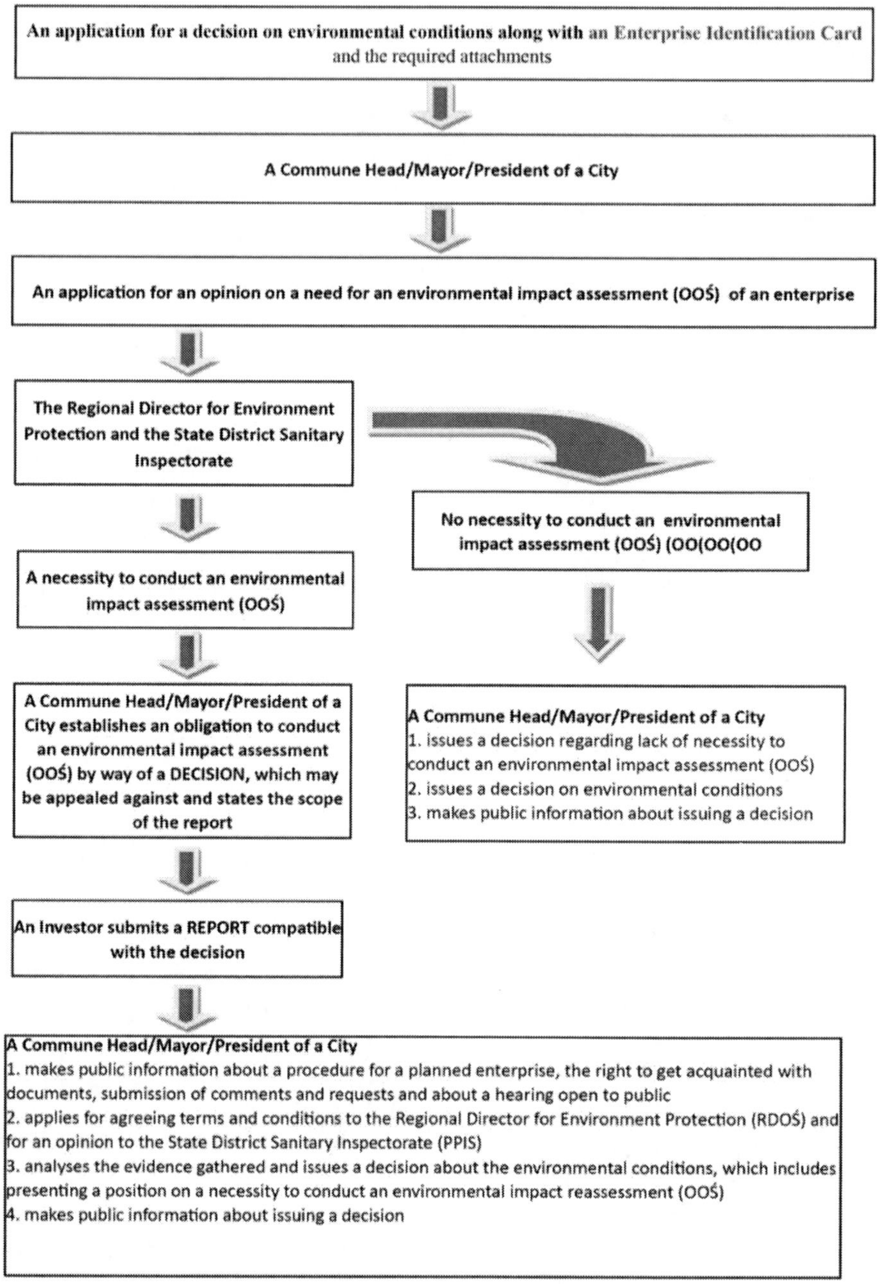

Fig. 1. The proceedings for issuing of the environmental decision for enterprises which could have significant environmental impact (own study based on [3]).

3 The research methodology

The SWOT method together with its variations is a universal tool and widely used, because it allows analyzing various research objects, i.e .:
- urban development planning - a strategy for the city of Dar es Salaam in Tanzania in his publication *A SWOT analysis of strategical urban development planning: The case of Dar es Salaam city in Tanzania* F. Halla [5];
- university development - an example is the University of Warwick, whose strategy was presented by Robert G. Dyson in the work *Strategic development and SWOT analysis at the University of Warwick* [6];
- teams and organizations - this is one of the most popular research area for scholars using SWOT analysis. In the paper by S. E. Jackson, A. Joshi and N. L. Erhardt, *Recent Research on Team and Organizational Diversity: SWOT Analysis and Implications* [7], it was used to assess the effects of diversity of jobs in teams and organizations;
- evaluation of various sectors of the economy - an example is the publication of *Regional Energy planning through SWOT analysis and strategic planning tools .: Impact on renewables development*, by J. Terrados, G. Almonacid, L. Hontoria [8] and *SWOT Analyzes of the National Energy sector for sustainable Energy development* presenting the results of research by the scientists Markovska N., Taseski V., Pop-Jordanova [9];
- development of companies, enterprises and products - presented in the following publications: *A knowledge-based SWOT - analysis system as an intrument for strategic planning in small and medium sized enterpises* [10] and *SWOT analysis: It's time for a product recall* [11].

SWOT analysis can be equally useful in use with other methods present in the article *Utilizing the analytic hierarchy process (AHP) in SWOT analysis — a hybrid method and its application to a forest-certification case* [12] or *Application of a quantification SWOT analytical method* [13].

SWOT analysis is based on a basic classification diagram of factors, which influence current and future condition of a researched structure. The factors can be classified into external and internal and those impacting in a positive or negative way. Mixing these two types of classifications generates four categories of factors:

- External positive- opportunities,

External positive factors, otherwise known as opportunities, are possibilities to use a situation in the surrounding of a researched structure in order to achieve the aim pursued. If they are identified early and correctly, they will constitute a development impetus, or they will allow to limit the risks.

- External negative- threats

Factors within the environment of a researched structure, namely threats, are barriers constituting danger of an adverse change taking place [14], which causes delays, incomplete aim achievement or its complete failure. It often leads to additional costs, efforts and outlays [14].

- Internal positive- strengths,

Strengths are the advantages of a researched structure causing an object in question to stand out, which allows to obtain an advantage.

- Internal positive- strengths,

Internal conditions (the features of a researched structure, which are dependent on it) constitute a weakness, a defect or some negative conditions effecting the building of an advantage (strength) or even constituting a barrier to this process [14].

The above mentioned classification can be presented in a schematic way (see Figure 2).

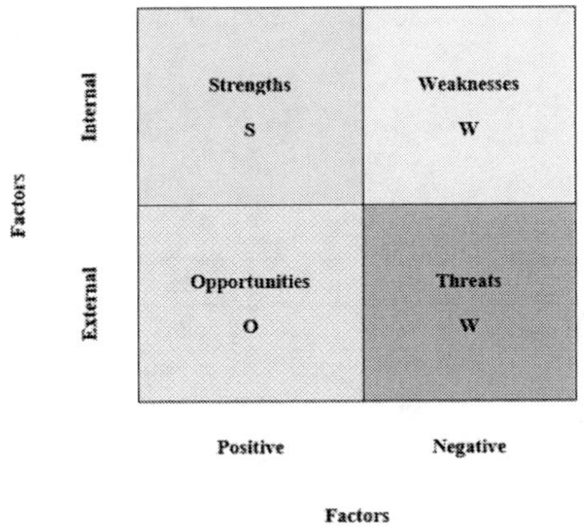

Fig. 2. The classification of factors in SWOT analysis [own study].

In the literature, many modifications of the basic concept of SWOT analysis can be specified. It is due to various approaches among the authors regarding the essence, the aim and the perspective of conducting the research. The most common modifications include WOT's- up analysis, TOWS analysis and TOWS-SWOT analysis [15].

The latter is a combination of TOWS analysis and SWOT analysis and it aims at conducting the analysis from the inside to the outside and from the outside to the inside. This approach allows not only to conduct a complex assessment of the weak and the strong points of the researched object and its opportunities and threats in the environment, but also to identify how the researched object impacts the environment and how the environment impacts the object.

General objectives of conducting TOWS-SWOT analysis consist of 10 steps presented below [own study based on [15]:

1. Indicating the aim of the analysis and characterising the diagnosed object.
2. Identifying as many as possible factors which impact the object and their characterising them.
3. Grouping the factors into sets S/W/O/T.
4. Classifying the factors as one of the three sub-groups (factors type A- relevant, factors type B- less relevant, factors type C- the least relevant).

* Corresponding author: emiurban@pg.edu.pl

5. Development of the potential strategic variants:
- aggressive strategy (SO) maxi-maxi,
- conservative strategy (ST) maxi- mini,
- competitive strategy (WO) mini-maxi,
- defensive strategy (WT) mini- mini.

6. Assigning weights to particular factors in the set in such a way that their sum is 1.
7. Using 8 cross tables to indicate correlation between the factors assigning them value 0, 1 or 2.
8. Performing calculations for each of the 8 cross tables and placing the results in the summary
9. Interpretation of the results based on the information from the summary table.
10. Choosing a suitable action strategy for a design/organization for the period of conducting the analysis.

4 TOWS-SWOT analysis as a tool for assessing a role of an EIC in a building process of water tourist investments

TOWS-SWOT analysis has been used in order to assess a role of an Enterprise Identification Card in a building process of water tourist investments. The strong and weak points, defined as the matters of elaboration and content of the document, that is the internal surroundings, have been identified. The external surroundings is considered as opportunities and threats, that is the procedures connected with KIP, legal acts, etc.

The factors classified to particular groups and their weights have been presented in Table 1. Having performed calculations for each of the 8 cross tables, the summary table has been developed – Table 2. The choice of a suitable strategy has been made, based on the results of the TOWS-SWOT analysis summarized in Table 3.

Table 1. An assessment of the strength of the influence of particular conditions.

Code	Internal factors		Code	External factors	
	Weight	The strong points		Weight	Opportunities
S1	0.3	KIP aims to specify the conditions necessary for an investment to be completed	O1	0.2	Employing more people to process KIP and issue decisions
S2	0.2	A necessity to describe the applied technology by an Investor allows to verify the possibility of the nature protection	O2	0.3	Establishing the maximum time of 2 months for the consideration of KIP and issuing a decision
S3	0.2	A necessity to state a type and anticipated quantity of substances or energy released to the environment when applying methods, which protect the environment	O3	0.2	Simplification of KIP preparation in the future
S4	0.1	The detailed nature of KIP	O4	0.1	Reduction of the submitted attachments
S5	0.2	A necessity to state the anticipated amount of water, resources, materials, fuels and energy used in the period of enterprise and exploitation	O5	0.1	Establishing one institution processing KIP in each voivodship
			O6	0.1	Taking over a necessity to indicate application of specific eco-friendly solutions by competent officials with technical knowledge
Total	1.0		Total	1.0	
Code	Internal factors		Code	External factors	
	Weight	The weak points		Weight	Threats
W1	0.2	Although at this stage the arrangements regarding a specific variant of an investment are on the way, in KIP all the variants of the enterprise must be defined	T1	0.1	A necessity for an investor to classify the enterprises which can potentially have a significant impact on the environment
W2	0.15	The numerous necessary attachments to KIP	T2	0.1	Common requests to supplement KIP
W3	0.25	A necessity for the applicant to refer to all possible forms of the environment protection	T3	0.1	Too high employment costs of a specialist environmental company for an applicant
W4	0.15	A difficulty with the correct preparation of KIP	T4	0.2	A complicated procedure
W5	0.25	A necessity to describe potential cross-border impacts on the environment by an applicant	T5	0.4	Delays or a lack of investment implementation, especially from the state budget grants or foreign funds
			T6	0.1	Lack of technical knowledge among the officials issuing an environmental decision
Total	1.0		Total	1.0	

Table 2. The summary of the results of the performed SWOT-TOWS analysis assessment of the strength of the influence of particular conditions.

Combination	Results of SWOT analysis		Results of TOWS analysis		Summary results of SWOT/TOWS	
	The sum of the interactions	The sum of the products	The sum of the interactions	The sum of the products	The sum of the interaction	The sum of the products
S/O	32	5.5	48	8.8	80	14.3
S/T	56	11.6	58	11.9	114	23.5
W/O	44	9.4	56	11.45	100	20.85
W/T	54	9.45	64	11.9	118	21.35

Table 3. A selection of a strategy based on the results received upon the analysis.

	Opportunities	**Threats**	
	O	**T**	
The strong points [S]	*Aggresive strategy*	*Conservative strategy*	
	80	114	The sum of number of interactions
	14.3	23.5	The sum of the weighted number of interactions
The weak points [W]	*Competitive strategy*	***Defensive strategy***	
	100	**118**	The sum of number of interactions
	20.85	**21.35**	The sum of the weighted number of interactions

In consequence of the performed calculations, the result obtained indicates that a role of an Enterprise Identification Card in a building process of water tourist investments is defensive. It is a situation where there are more threats than opportunities and the weak aspects dominate over the strong ones in the external surroundings. That means that KIP is a complicated and time - consuming document to prepare. It requires attaching many documents, which in most cases an applicant must acquire at a charge, despite the fact that the institution dealing with KIP has free access to this information. The identification of the alternative variants for an enterprise is rather illusory because most of the arrangements required by law and the construction design itself is virtually ready at this stage. Individuals presenting the alternative solutions do not concentrate on proposing better or equal design variants, but they present the worse ones so the variant chosen by the Investor is the better one for the institution considering the submitted KIP. The necessity to refer to all possible forms of the environmental protection and potential transboundary impacts on the environment is difficult to formulate for a person preparing a project due to number of data and lack of specialist knowledge. In most cases, an applicant cannot afford to employ specialists to formulate KIP. Unfortunately, companies specializing in development of documents and environmental analysis are often on the side of the authority issuing an environmental decision. Moreover, these services are not cheap and their cost is often not included in execution of tenders [16]. As a consequence, an applicant prepares

KIP on his or her own. A problem appears when there is a necessity to classify an enterprise as the one which can potentially impact the environment or can impact the environment in a significant way. However, the lack of knowledge on the part of an applicant is not the only reason for problems with drawing up of KIP. The lack of professional technical knowledge on the scope of construction among the officials causes misunderstandings and complications in the procedure of issuing an environmental decision. The frequent calls upon supplements or explanations and lack of possibility to achieve a consensus between an applicant and an official very often results in a few months long delays in an investment process, or in the worst cases, in suspension of the completion of an investment and loss of grants from the state budget or foreign funds.

5 Conclusions

An Enterprise Identification Card (KIP) is a document required by the law system of the environmental protection in Poland for the enterprises which could have a significant impact on the environment. Most water tourist investments are included in this category. KIP is supposed to specify the conditions under which a building construction can be completed. However, as the TOWS-SWOT analysis has shown, a role of an Enterprise Identification Card has a defensive character. It means that instead of facilitating the investment process, it complicates it to a significant extent. The reason is the fact that everybody can prepare KIP, hence the main infringements are on the part of the applicants already at a preliminary stage, related to the formal requirements and the classification of an enterprise. Each infringement in documents not only contributes to the increase of waiting time for a decision, but it also raises the costs of an enterprise or its suspension [4].

According to the authors, the result of the analysis should be the reason for a debate on simplification of not only the procedure of granting the environmental decision itself, but also on delegating the necessity of formulating the guidelines for the environmental protection in a given area in an investment process to the officials, who possess greater knowledge in this regard than an applicant. That of course means an increasing the number of employees who would verify the content of KIP, but it could also considerably reduce the waiting time for an environmental decision. In consequence, the progress of a building process, including inter alia water tourist investments would be significantly improved.

References

1. The Council of Ministers Regulation of 9 November 2010 on the enterprises, which could have a significant environmental impact. Journal of Laws 2010, No.213, item 1397 as amended.
2. The Council Directive of 27 June 1985, No 85/337/EWG about an assessment of the impact of some enterprises on public and private environment
3. P. Wójcik, *Decyzja o środowiskowych uwarunkowaniach oraz ocean oddziaływania przedsięwzięcia na środowisko*, Poradnik dla Inwestora, access: 15 December 2017, http://www.korfantow.pl/download/attachment/28630/decyzja-o-srodowiskowych-uwarunkowaniach-oraz-ocena-oddzialywania-przedsiewziecia-na-srodowisko-poradnik-dla-inwestora.doc, (2015) (in polish)
4. L. Korona, *Uwarunkowania realizacji planowanych przedsięwzięć budowlanych z uwzględnieniem wymogów ochrony środowiska*, Zeszyty Naukowe Politechniki Gdańskiej, Budownictwo Lądowe, **61**, 15-22 (2007)
5. F. Halla, *A SWOT analysis of strategic urban development planning: The case of Dar es Salaam city in Tanzania*, Habitat International, **31**, 130-142, (2007)
6. R. G. Dyson, *Strategic development and SWOT analysis at the University of Warwick*, European Journal of Operational Research, **152(3)**, 631-640, (2004)
7. S. E. Jackson, A. Joshi, N.L. Erhardt, Recent Research on Team and Organizational Diversity: SWOT Analysis and Implications, Journal of Management, **29(6)**, 201-830, (2003)
8. J. Terrados, G. Almonacid, L. Hontoria, Regional energy planning through SWOT analysis and strategic planning tools.: Impact on renewables development, Renewable Susainable Energy Reviews, Elsevier, **11(6)**, 1275-1287, (2007)
9. N. Markovska, V.Taseska-Gjorgievska, J. pop-Jordanov, SWOT analyses of the national energy sector for sustainable energy development, Energy, **34(6)**, 752-756, (2009)
10. G. Houben, K. Lenie, K. Vanhoof, A knowledge-based SWOT-analysis system as an instrument for strategic planning in small and medium sized enterprises, Decision Support Systems, **26(2)**, 125-135, (1999)
11. T. Hill, R. Westbrook, SWOT analysis: It's time for a product recall, Long Range Planning, **30(1)**, 46-52, (1997)
12. M. Kurttila, M. Pesonen, J. Kangas, M. Kajanus, *Utilizing the analytic hierarchy process (AHP) in SWOT analysis - a hybrid method and its application to a forest-certification case*, Forest Policy and Economics, **1**, 41-52, (2000)
13. H.H. Chang, W.Ch. Huang, *Application of a quantification SWOT analytical Method*, Mathematical and Computer Modeling, **43**, 158-169, (2006)
14. W. R. Griffin, *Podstawy zarządzania organizacjami*, Wydawnictwo Naukowe PWN, Warsaw, 250-253 (2004) (in polish)
15. M. Nowicki, SWOT, scientific editing: K. Szymańska, *Kompendium metod i technik zarzadzania. Teoria i ćwiczenia*, Printing house a Wolters Kluwer business, Warsaw, 325-354 (2015) (in polish)
16. B. Grzyl, E. Miszewska-Urbańska, M. Apollo, *The life cycle cost of a building from the point of view of environmental criteria of selecting the most beneficial offer in the area of competitive tendering*, E3S Web of Conferences, **17**, 00028 (2017)

Gamma radiation as a source of information about the characteristics of celestial bodies

Marek Przyborski[1], *Zuzanna* Siemieniago[1,*]

[1]Gdańsk University of Technology, Faculty of Civil and Environmental Engineering, Gdansk, Poland

Abstract. The Universe is a violent and hostile environmental. Remote sensing methods give the possibility to search for the answer to the most fundamental question that humankind has ask, how the universe works? In this paper we would like to take a closer look at the some specific objects in the Universe that their influence on its evolution is still a matter of research all over the world. The Gamma Bursts are one of the most dangerous phenomenon that we have ever discovered. Astronomers may observe them also when watching and searching for another form of energy propagated through the time and space gravitational waves caused by binary systems of neutron stars. In the following we would like to explain the fundamentals of multi-messenger astronomy.

1 Introduction

Gamma ray bursts accompany one of the most dramatic events that occur in the universe. The observation of gamma radiation allows the discovery of further celestial bodies and the study of already known ones like pulsars, black holes, nuclei of active galaxies or dark matter [12, 16, 21, 3]. In space, there are constantly many processes emitting huge amounts of energy, such as star bursts or asteroid collisions that cause weak gamma-ray bursts [12]. Electromagnetic waves accompanying events occurring in the Universe arise in microscopic processes, strongly interact with the surrounding matter, are absorbed and dispersed. The information they provide relates to the moment of the last contact with matter, e.g. the surface of the last scattering, which may be the surface of the star [3].

Gamma bursts can be divided, in terms of duration, into short and long [21]. Long, lasting from several seconds to even several minutes, called hibernate, accompanied by powerful supernova explosions [14]. Short in turn, they last only a fraction of a second and result from the merging of neutron stars-objects created after supernova explosions [17].

Neutron stars are objects with small diameters, not exceeding 25 km, but they have very high density of matter. Bearing in mind the principle of conservation of the angular momentum and the characteristics of the object, it is not surprising that its incredible speed of rotation, which reaches even more than 700 revolutions per second. They were created as a result of the collapse of the nucleus of a star with a large mass (8-10 solar masses). Such a phenomenon brings about a rapid increase in the magnetic field, hence neutron stars have a magnetic field that is at least one hundred million times more powerful than the Earth's magnetic field [12]. Such a magnetized, fast-rotating star, hereinafter referred to as a pulsar, generates electricity, which is associated with the emission of electromagnetic radiation [13]. This is most often radio or X-ray radiation.

The discovery of pulsars in the 1960s gave rise to their further observations and research. Soon a pulsar was identified, called the Hulse-Taylor pulsar, with a variable period, from the names of the discoverers. It turned out to be part of the double star system. The general theory of relativity shows that accelerating masses should produce gravitational waves carrying quanta of energy, so that stars circling in orbits lose energy and the diameters of orbits decrease. This results in the stars' constant approaching each other and finally their collision. The final effect of such a collision will be the creation of a black hole. The material released as a result of the collision is ejected into space at a considerable speed. It reaches the border of the dual system, where it meets interstellar matter. Then, shock waves create electromagnetic radiation [5].

Currently, the effect of the above events is also observed as gamma radiation. Unfortunately, the flare itself does not allow for precise positioning of its source and distance. Also the very nature of gamma radiation makes it difficult to find the location of the source. The emission of radiation flux looks different depending on the case. During long gamma-ray bursts, the radiation is emitted along the axis of rotation of the collapsing star. The phenomenon takes place differently in the situation of the collision of neutron stars, then the emission occurs in the direction perpendicular to the orbital plane of the connecting objects. Looking along the beams we will see a strong flare, but if the detector is directed at a different angle, it may not register the signal at all.

Very often, the accompanying phenomenon is the presence of X-rays, which have only a little less energy

* Corresponding author: z.siemieniago@gmail.com

than gamma, the emission of which is longer, and allows a more accurate determination of the location of the source of the flash [21]. The glow of X - ray and visible light seen on Earth connected with the flash is visible even above one day, which in comparison with the duration of a short gamma ray gives a better chance of observing the phenomenon.

Gamma radiation is not directly recorded from Earth [1]. The atmosphere that absorbs some ranges of wavelengths, e.g. lethal to cells of living organisms - gamma rays.

There are indirect methods, e.g. Cherenkov radiation registration, which is emitted as a result of the interaction of gamma photons with the Earth's atmosphere. Said radiation is emitted when the charged particle moves at a speed greater than the phase velocity of light in the given medium, in a precisely defined direction at an acute angle to the direction of motion of the particle [2].

The technology described is used by projects such as CTA (Cherenkov Telescope Array) or VERITAS (Very Energetic Radiation Imaging Telescope Array System) [17]. Both are equipped with terrestrial telescopes with a mirror of 4 to 24 meters. In these, and similar projects, sensitive charged-coupled device (CCD) cameras or multi-pixel photomultiplier tubes (abbreviated as PMT) are used for registration.

Observations should meet several conditions to identify the gamma-ray burst:
- have the same intensity on each of the telescopes,
- appear in the same place in each camera,
- be a point phenomenon.

Such assumptions allow to exclude the possibility of observing the background radiation, not the gamma ray [14].

It should be emphasised that Cherenkov radiation is scattered relative to the gamma ray beam. It is usually registered as a several-kilometre area. The centroid of this field should represent the source of the emitted wave [4].

A lot of information is provided by the international space gamma ray observatory (Gamma-ray Large Area Space Telescope) today called the Fermi Gamma-ray Space Telescope from the name of the professor, pioneer in high energy physics, Enrico Fermi. Launched in 2008, the Fermi telescope carries out continuous observations of the sky in the field of high energy waves. It is equipped with two instruments:
- LAT - Large Area Telescope - detector with high sensitivity and resolution, covers about 1/5 of the whole sky,
- GBM - Gamma-ray Burst Monitor - a supplementary device for the LAT, has a very short response time, which allows the registration of gamma-ray bursts [20].

The work of both instruments differs from each other. When the gamma ray gets to the LAT, it affects one of the 16 tungsten sheets. This interaction transforms the gamma ray into electron and positron, which allow you to follow the movement. Thanks to such transformation, LAT is able to determine the direction of radiation. GBM detectors capture X-rays or low-energy gamma rays. At the moment when the rays fall on one of the 12 detectors made of sodium iodide, a flare of the registered photomultiplier occurs. By comparing the signal speed of at least 4 GBM detectors it is able to triangulate to determine the position of the source of the burst.

The combination of two instruments has allowed many discoveries, including pulsars, identification of the largest gamma-ray bursts, background radiation sources [13]. A significant part of the data from both the LAT and GBM observations is available on the Internet on the website: https://fermi.gsfc.nasa.gov/.

2 Collisions of neutron stars

The collisions of neutron stars are phenomena which, according to Einstein's theory, should be accompanied by the appearance of gravitational waves. The radiation generated during this explosion, in addition to radio ranges, should also be supplemented with the visible range. The observation of this event in the visible range and observations in the field of gamma and X-ray radiation are a kind of warning signal, which proves that we can also deal with the emission of gravitational wave [5]. Thanks to observations conducted in various ranges, we are able to identify the source of the gravitational wave. In this way a new type of astronomy was created which was initiated with the creation of the Laser Interferometer Gravitational-Wave Observatory (LIGO) project [10]. One of the last events recorded by LIGO observatories probably occurred at a distance of 130 million light years from Earth, in the galaxy NGC 4993. The distance indicates that astronomers have to deal with the closest, ever observed, source of gravitational waves that have been recorded [11]. It is also one of the closest registered gamma-ray bursts. According to the researchers, newly detected phenomena is also signalling the presence of so-called kilonova - an object whose existence theorists have been predicted for over 30 years. Kilonova can be the main objects responsible for the proliferation of very heavy particles in space, including gold and platinum [6]. According to the researchers, simultaneous detection of gravitational waves and their electromagnetic counterpart (including visible light) from the same source marks the beginning of a new era in astronomy, which they describe as "multi-messenger astronomy" [11, 18]. The signal of gravitational waves (event marked GW170817 its localization is presented in Figure 1) was registered on 17 August 2017 at 14:41 CEST is a twin detector LIGO, located in Hanford, Washington, and Livingston, Louisiana (USA) [11].

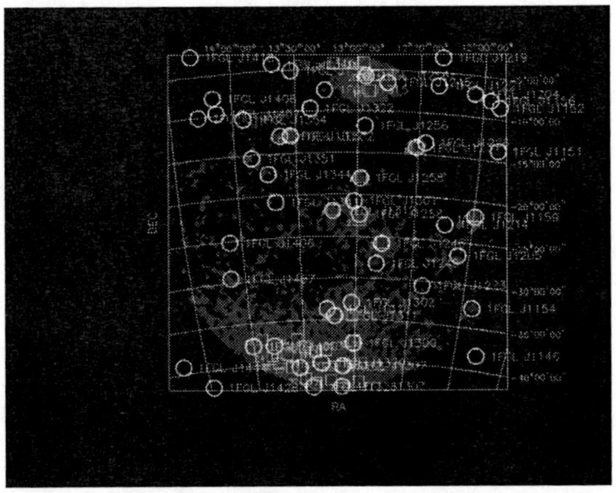

Fig. 1. Location of GW170817 (source: http://cds.u-strasbg.fr).

The participation of the Virgo Observatory allowed for the specification of the sky area from which the signal came. Then, the work involved observatories designed to study electromagnetic waves in various wavelength ranges, combing the sky in search of an electromagnetic equivalent for the source of gravitational waves [15, 19]. Characteristic of the so-called gravitational waves, in this case, the tweet lasted almost two minutes and the last 30 seconds are shown in the spectrogram below Fig. 2.

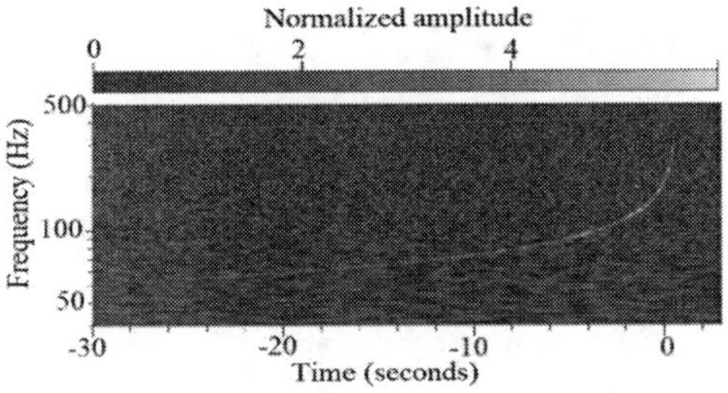

Fig. 2. Spectrogram of signal CHIRP (source: www.ligo.org/).

3 Spectrogram

Detailed analysis of this signal indicates that the observed system is much less massive, and the mass of the system components correspond to the double system of neutron stars [18]. The possibility of observing the evolution of such a system in the range other than the spectral ranges used so far, gives the possibility of a thorough analysis of one of the most extreme objects found in the Universe. The fact that these objects are created during the explosion of supernova stars, where the matter collapses to a density many times greater than the density of atomic nuclei and we are not able to reproduce this process on Earth, causes that the data collected during the observation of the evolution of such a dual system is very unique [17]. Registered by the Fermi satellite, simultaneous with the collapse, gamma-ray burst, was named GRB 170817A [7]. The characteristic curve of the radiance changes from this source agrees with theoretical models of the Kilonova, or clouds of neutron-sized matter remaining after the collision of neutron stars, which shines due to radioactive decays of elements. This event made it possible to test the correctness of the new method of measuring distances in the Universe, independent of the previously used so-called methods of "standard candles". The observed double system can be seen as a "standard siren" [18] providing a distance that it compares with the known escape velocity of the NGC 4993 galaxy, ϑ_H, i.e. using Hubble's law:

$$\vartheta_H = H_0 d \quad (1)$$

where :

ϑ_H- recessional velocity, the velocity of galaxy getting away, expressed in km/s;

H_0 - the constant of proportionality, Hubble constant expressed in km/s/Mpc:

$$H_0 = 70^{+12}_{-8} km \cdot s^{-1} Mpc^{-1} \quad (2)$$

d- distance to the galaxy masured in Mpc.

Thus, in this way, we have also independent from other methods of measurement of the Hubble constant, and thus a new estimation of the rate of Universe spread, which is compatible with measurements from supernovae and microwave background radiation measurements. Thanks to the measurements made during the event GW 170817, a comparison of the time shift between gravitational waves and photons was also made, which allowed to estimate the speed of gravitational waves, which turns out to be almost identical to the speed of light.

During the observation of the GW 170817 event, scientists launched one of the largest ever target observation campaigns, seeking the equivalent of a source of gravitational waves in many ranges of electromagnetic wavelength. As part of this action, one of the first, the optical equivalent of the source of gravitational waves was noticed by the infrared VISTA (Visible & Infrared Survey Telescope for Astronomy) telescope. A number of other European Southern Observatory (hereinafter ESO) telescopes were used, such as the Very Large Telescope (VLT), the VLT Survey Telescope (VST), the New Technology Telescope (NTT), as well as the Atacama Large Millimeter Array (ALMA) radio telescope network and other partner instruments working in the ESO observatories in Chile. These instruments followed the object for subsequent days and weeks after detection. The observations were also conducted from space, e.g. by the Hubble Space Telescope. The NGC 4993 galaxy, where neutron stars joined, was observed by the OGLE team led by prof. Andrzej Udalski. The Polish project Pi of the Sky, implemented by the Faculty of Physics of the University of Warsaw, the National Centre for Nuclear Research of the Polish Academy of Sciences and the Centre of Theoretical Physics of the Polish Academy of Sciences, whose main goal is to search for optical flashes accompanying gamma flares.

4 Big Bang and LIGO

Thanks to the LIGO detectors, we have also been able to study the earliest moments of the Universe. The basis for these analyses are earlier discoveries of gravitational waves, the source of which was the collision of black holes. The gravitational waves discovered in the Advanced LIGO experiment, created by the double system of black holes, give one more opportunity to learn what the beginning of the universe looked like and allow to reach even further into the past. As assumed in the modern version of the Big Bang theory, at the very beginning the Universe underwent a phase of rapid expansion, so-called inflation, during which the original gravitational waves were generated. Like the relic of neutrinos and electromagnetic radiation, they fill the Universe today [10]. Unfortunately, the amplitude of the primary waves with frequencies measured in the Advanced LIGO experiment is much smaller than the sensitivity of this instrument. The waves of sizes comparable to the size of the observable universe have much greater amplitude, but to detect them, we would need such a detector size. The primary gravitational waves, however, have an effect on the polarity of the background radiation, so the radiation can act as a specific detector. The background radiation is partially polarized as a result of Thomson's dispersion on free charged particles, primarily electrons. However, for the scattered photons to be polarized, the electron radiation falling on the electron must have such a distribution that the directions corresponding to the maximum and minimum radiation intensity are set at right angles to each other. The gravitational waves passing through the ionized medium allow to generate such a distribution.

The directions of radiation polarization projected onto the observer's blue sphere will create a zero-source field. By analogy with the magnetic field in electromagnetism this type of polarization is called the B mode, in contrast to the E modes generated primarily by ordinary scalar disturbances. Detection of B mode polarization of relict radiation would be indirect evidence of the existence of primary gravitational waves and confirmation of the epoch of cosmological inflation in the early Universe. Thanks to that, we could find out what the Universe looked like only about a 10^{-36} second after the initial singularity [8, 9].

The detection of modes B polarization of background radiation is one of the most important goals of present and planned experiments on relict radiation. In 2014, such a detection was reported by the BICEP2 experiment team [8, 9]. However, thanks to the data from the Planck satellite, it quickly turned out that the observed B mods are related to the polarization of radiation emitted by our Galaxy. So the search for traces of the original gravitational waves is still going on. However, the lack of detection already allows to estimate the upper limit of energy at which inflation could occur. This boundary is about 10^{16} GeV and so, there are several orders of magnitude lower than the Planck energy, at which gravity separated from the other interactions.

The answer to the question about even earlier moments of the Universe depends on what happened before the inflation era. It can be assumed that with the already mentioned Planck energy some physical processes have occurred, the traces of which can be preserved in the properties of relict radiation, neutrino background or primary gravitational waves. However, to better understand this, we need a quantum theory of gravity describing the interaction with the Planck energy. It is worth mentioning that according to some theories aspiring to the quantum theory of gravity in the era of such a high energy density loses the meaning of the notion of time known to us, and the smallest unit of time, having a physical meaning, is about a 10^{-44} second.

5 Summary

The events presented in this paper, followed, among others, by LIGO observatories, are an example of research in the new era in astronomy. Multi-messenger astronomy gives us a wide range of research possibilities of the Universe. Until now, astronomy was associated with long-lasting phenomena, because such observations could be observed in the spectral ranges. Observations of gravitational waves give the possibility of recording phenomena lasting very shortly, in the order of seconds. The widespread availability of these data gives researchers the opportunity to improve their methods, improve software codes, and understand the phenomenon that Albert Einstein predicted 100 years ago.

References

1. W. Romanishin, *An Introduction to Astronomical Photometry Using CCDs. Image (Rochester, N.Y.)*, 167. (2002)
2. J. B. Kepka, *IV.5. Promieniowanie Czerenkowa*, 2–6 (1984) (in Polish)
3. A. Subramaniam, R. Sagar, *Multicolor CCD Photometry and Stellar Evolutionary Analysis of NGC 1907, NGC 1912, NGC 2383, NGC 2384, and NGC 6709 Using Synthetic Color-Magnitude Diagrams*. The Astronomical Journal, **117, 2**, 937–961 pp. (1999) DOI:10.1086/300716
4. J. Aleksić, J. Ric, M. Martínez, *Optimized analysis method for indirect dark matter searches with Imaging Air Cherenkov Telescopes*. Proceedings of Science. (2013) DOI:10.1088/1475-7516/2012/10/032
5. J. Centrella, *Gravitational wave astrophysics: Opening the new frontier.* AIP Conference Proceedings, **1381**, pp. 98–116 (2011) DOI:10.1063/1.3635827
6. B.P. Abbott, R. Abbott, T.D. Abbott, F. Acernese, K. Ackley, C. Adams, et al., *A gravitational-wave standard siren measurement of the Hubble constant.* Nature, **551**, 7678, 85–98 pp. (2017) DOI :10.1038/nature24471
7. S.J. Smartt, T.W. Chen, A. Jerkstrand, M. Coughlin, E. Kankare, S.A. Sim, et al., *A kilonova as the electromagnetic counterpart to a gravitational-wave*

source. Nature, **551**, 7678, 75–79 pp. (2017) DOI :10.1038/nature24303

8. A. Goldstein, P. Veres, E. Burns, M.S. Briggs, R. Hamburg, D. Kocevski, et al., *An Ordinary Short Gamma-Ray Burst with Extraordinary Implications: Fermi-GBM Detection of GRB 170817A*. (2017) DOI:10.3847/2041-8213/aa8f41
9. P.A.R. Ade, R.W. Aikin, D. Barkats, S.J. Benton, C.A. Bischoff, J.J. Bock, et al., *Detection of B -mode polarization at degree angular scales by BICEP2.* Physical Review Letters, **112**, 24 (2014) DOI:10.1103/PhysRevLett.112.241101
10. P.A.R. Ade, N. Aghanim, Z. Ahmed, R.W. Aikin, K.D. Alexander, M. Arnaud, et al., *Joint Analysis of BICEP2/Keck Array and Planck Data*. Physical Review Letters, **114**, 10, 1–17 pp. (2015) DOI:10.1103/PhysRevLett.114.101301
11. S. S. Kimura, K. Murase, I. Bartos, K. Ioka, et al., *Trans-Ejecta High-Energy Neutrino Emission from Binary Neutron Star Mergers* (2018) arxiv.org/abs/1805.11613v1
12. S. Kaufmann, O. Tibolla, *Ancient Pulsar Wind Nebulae as a natural explanation for unidentified gamma-ray sources.* (2018) arxiv.org/abs/1805.11522v1
13. R. T. Bartels, T. D. P. Edwards, C. Weniger, *Bayesian Model Comparison and Analysis of the Galactic Disk Population of Gamma-Ray Millisecond Pulsars.* (2018) arxiv.org/abs/1805.11097
14. Y. Mizumura, A. Takada, T. Tanimori, *Systematic Study of Nuclear Gamma-Ray Spectra of One Hundred Super Novae Expected by Future Nuclear Gamma-Ray Imaging Spectroscopic Observations.* (2018) arxiv.org/abs/1805.07939
15. V. Tatischeff, A. De Angelis, M. Tavani, I. Grenier, et al., *The e-ASTROGAM gamma-ray space observatory for the multimessenger astronomy of the 2030s.* (2018) arxiv.org/abs/1805.06435
16. H.E.S.S. Collaboration, H. Abdallah, A. Abramowski, F. Aharonian et al., *Search for γ-ray line signals from dark matter annihilations in the inner Galactic halo from ten years of observations with H.E.S.S.* (2018) arxiv.org/abs/1805.05741
17. S. Celli, G. Morlino, S. Gabici, F. Aharonian, *Supernova remnants in clumpy media: particle propagation and gamma-ray emission.* (2018) arxiv.org/abs/1804.10579
18. J. Wei, X. Wu, H. Gao, *Cosmology with Gravitational Wave/Fast Radio Burst Associations.* (2018) arxiv.org/abs/1805.12265
19. The LIGO Scientific Collaboration, the Virgo Collaboration: B. P. Abbott, R. Abbott, T. D. Abbott, et al., *GW170817: Measurements of neutron star radii and equation of state.* (2018) arxiv.org/abs/1805.11581
20. G. Wang, Z. Wang, *Searching for γ-ray Counterparts to Very Faint X-Ray Transient Neutron Star binaries.* (2018) arxiv.org/abs/1804.09871
21. V. Liccardo, M. Malheiro, M. S. Hussein, T. Frederico, *Nuclear processes in Astrophysics: Recent progress.* (2018) arxiv.org/abs/1805.10183

E3S Web of Conferences **63**, 00003 (2018)
2018 BGC
https://doi.org/10.1051/e3sconf/20186300003

Analysis and design improvement of high class roads' arched sections in Poland

Joanna Żukowska[1,*], *Paweł* Słomiński[1]

[1]Gdańsk University of Technology, Faculty of Civil and Environmental Engineering, 11 Narutowicza Street, 80-213 Gdańsk, Poland

Abstract. Although Polish road network standards, in particular higher classes of roads are improving dynamically the number of fatalities on the network is still higher that EU's average. Ensuring road safety and high maintaining standards is a complex issue and one that poses a major challenge for all road authorities. Road accidents at high speed roads characterize specially serious material and social consequences, thus research into road safety is of a special interest and numerous analyses and debates are being conducted worldwide as well as in Poland. In this paper the authors provide the results and recommendations coming from the analysis of accidents which happened on high speed roads during the heavy rainfall. The analysis of incidents which happened on sections with changing vertical alignment shows that a properly designed road geometry does not eliminate road accidents or conflicts.

1 Introduction

Ensuring road safety and high maintaining standards is a complex issue and one that poses a major challenge for all road authorities [1]. Road accidents at high speed roads characterise specially serious material and social consequences, thus research into road safety is of a special interest and numerous analyses and debates are being conducted worldwide as well as in Poland. Globally motorway systems generate nearly 50 million persons injured in traffic accidents every year, not including the 1.2 million deaths [2]. According to Police statistics traffic injury is the single leading cause of unintentional death in the first 5 decades of human life.

Traffic safety management is a systematic process that aims to reduce the occurrence and severity of traffic accidents. The interaction between a man and a car within road traffic systems is unstable and poses a great challenge to highway safety management. The key for increasing the safety of highway systems is to design, build, and maintain them to be far more tolerant of the average range of this man/car interaction with motorways. Technological advancements in highway engineering have improved the design, construction, and maintenance methods used over the years. These advancements have allowed for newer highway safety innovations. One of them is being considered within this paper. The design stage is crucial for achieving the most appropriate location, alignment, and shape of a motorway. The design process involves the consideration of three major factors: human, vehicular, and roadway- and how these factors interact to provide a safe road [3]. Human factors include e.g. reaction time for braking and steering, visual acuity for traffic signs and signals, and car-following behaviour. Vehicle considerations include vehicle shape and dynamics that are essential for determining lane width and maximum slopes, and for the selection of design vehicles.

2 Road geometry and safety

Geometric design aims at optimizing the efficiency and safety while minimizing cost and environmental impacts. The process of geometric motorway design can be broken into three main parts: design of the alignment, profile, and cross-section of the road. Combined, the three elements provide a three-dimensional layout for a roadway. The alignment is the track of the road, defined as a series of horizontal tangents and curves. The profile is the vertical aspect of the roadway, including crest and sag curves, and the straight grade lines connecting them. Last bit now least the cross-section shows the position and number of vehicle and bicycle lanes and well as sidewalks, along with their cross slope or banking. Cross sections also show drainage features which are of the special interest of this paper as the element of design which, especially on a high speed roads, may strongly influence safety level [4].

The problem of water on the road surface is as follow: after falling onto road surfaces, rainfall runoff drains to the lowest point and in moving across the road surface forms a layer of water of varying thickness. This water can generate special problems to the road users. Splash and heavy spray are thrown up by moving cars, reducing visibility, while the water on the pavement reduces friction between the tires of the vehicles and road surface. Excessive water on the pavement, whether ponded or flowing, can represent a real risk of aquaplaning or the build-up of a layer of water between the vehicle tire and the road surface, which may lead to a

* Corresponding author: joanna.zukowska@pg.edu.pl

total loss of grip. While part of road surface drainage, aquaplaning is a critically important aspect of road surface drainage.

Accidents tend to be more frequent on road sections where a sudden change in road character, maintenance standards or surrounding conditions violates the driver's expectations. Aquaplaning may be one of these circumstances. A thin layer of water between the wheels of the car and the road surface may lead to a loss of traction that prevents the car from responding to control inputs. If it occurs to all wheels simultaneously, the car becomes an uncontrolled sled.

In Polish road design guidelines water on the roadway and how it is drained off is addressed in particular in relation to vehicles, traffic and tire grip under special conditions, while in other countries also the official instructions show the need to consider of changing vertical alignments of the road. Some analysis were conducted worldwide and the synthesis of their results is provided in the article. In this paper the authors present some construction methods to improve those road sections which were designed properly but still during the rainfall the accidents appear on them. In fact only two countries introduced guidelines on how to design the construction. These are Germany and Sweden. The guidelines of the Supreme Building Office of Bavaria's Ministry of the Interior, "Safe street design" describe how rainfall and driver behavior contribute to road accidents and what are the ways to design and re-design roads to increase safety [5]. Sweden's Road and Street Design Guidelines *"Råd för Vägars och gators utformning"* offer the so called "airscrew" solution where there is a gradient to be adjusted [6]. This may be applied on roads with one or more lanes. On November 19th 2008 the EU Parliament announced Directive 2008/96 / EC on road infrastructure safety, which obligate the Member States to introduce safety management rules on European transport network. Unfortunately, in Poland the legislative changes to the present are not in force on the whole road network. Thus, at least within the proper design and maintenance of sections with changing vertical alignment it is still a good moment to gain experiences and recommend best practices [7], [8].

3 Review of best practices

Tackling with the problem of water on the roadway several practices has been introduced in Poland and abroad. There are some of them introduced in this section.

3.1 Use of dedicated vertical signs

Special caution is advised when using sites with poor drainage during rainfall. This is due to poor workmanship, wrong interpretation of technical conditions for toll motorways and changing cross slopes. An experimental warning sign is proposed (Fig. 1).

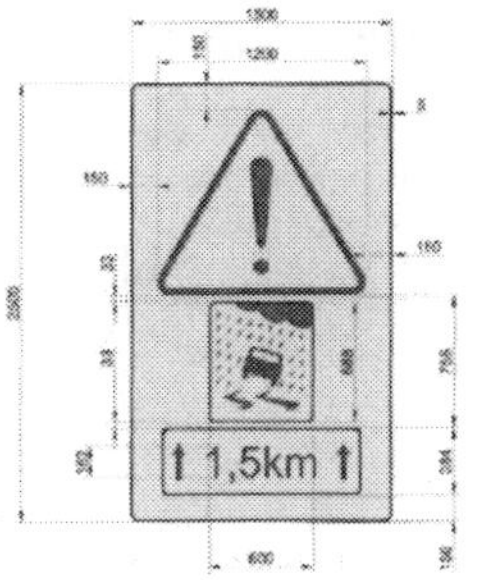

Fig. 1. Experimental signage to "inform about deteriorated conditions during rainfall"

3.2 Milling

Surface milling machines can be used to scarify the wearing course in order to correct water drainage. This is done by cutting off the near surface of the wearing course at several or about a dozen millimeters deep. Surface milling is an invasive method which cuts into the wearing course, opens up the pores between the aggregate's grains and polishes them. Milling can also make riding uncomfortable for two wheelers or even cause vehicles to lose steerability which in turn deteriorates road safety. Once uncovered, the aggregate is subjected to penetration of water, traffic and variable air temperature. As a consequence, the loss of binder may cause the grains to detach affecting the mineral and asphalt mix.

The work can be done using a Chinese milling machine *Masalta model MC8-4*, with a 5.5 KM Honda engine, working width of 200 mm and 20 mm milling depth. The machine is mainly used for scarifying horizontal road markings and treating the road surface. The next product which is available in Poland is the FA-300 milling machine manufactured by *HYDROG Zakład Budowy Maszyn* from Łódź. When the milling is only a few millimeters deep, the results during heavy rainfall have hardly any effect at all. The cost of the work including provisions to protect the site, restricted access to the road and mediocre drainage performance make it a short-term solution and a fairly costly one as well.

3.3 Proper vertical alignment

An analysis of incidents which happened during rainfall on sites with changing vertical alignment shows that a properly designed road geometry does not eliminate road incidents or conflicts. This problem is also covered in an analysis conducted by road safety experts. In addition, a road profile which meets the regulations will cost nearly just as much in road works costs as the "envelope" solution, but the impact on road safety is fairly good. The total cost is app. PLN 750,000.

 https://doi.org/10.1051/e3sconf/20186300003

3.4 Airscrew and envelope solution

By analyzing worldwide drainage solutions in places of a changing vertical alignment the authors has reached for technical regulations that are being used in the majority of European countries, Russia, the United States and Canada. From the analyzed eleven countries only two have guidelines for the "envelope" or "airscrew" solution. They were Swedish and German guidelines. It is not just coincidence that these countries are the most advanced in road safety.

4 Swedish guidelines

Swedish road and street design guidelines "Råd för Vägars och gators utformning" include a so called "airscrew" solution (Fig. 2) where there is a gradient to be adjusted. This is applied on one way roads with one or more lanes. In other cases the airscrew solution should be applied on an existing drainage for more comfortable driving. For any road width, the axis of "airscrew" rotation depends on the minimal length as illustrated in the formula below:

$$U = \Delta E \,/\, \omega \, * \, (VR + 10)/360 \qquad (1)$$

where: U – length of inclination (m), ΔE % - difference in gradient (%), ω – change of angle of rotation per unit of time, ≤ 0.038, VR - reference speed (km/h).

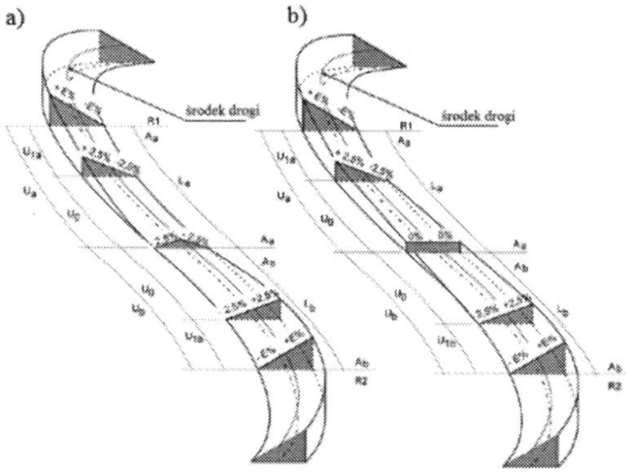

Fig. 2. "Airscrew" design: a) classic airscrew, b) airscrew with shifted zero points [6] ("srodek drogi" (pol.) = axe of the road).

5 German guidelines

The guidelines of the Supreme Building Office of Bavaria's Ministry of the Interior, "Safe street design" (Oberste Baubehoerde im Bayerischen Staatsministerium des Innern, Verkehrssichere strassengestaltung) describe how rainfall and driver behavior contribute to road accidents. Water determines grip between the tire and road surface. When turning the wheel or braking, the risk of skidding increases. The document also includes recommendations presented as a set of rules [5]. Where slopes are oppositely directed on roads, the minimum gradient along the road should be s≥ 0.7% or even s≥ 1.0%. This solution is designed to prevent water being drained from lower zones. To ensure sufficient road drainage, the guidelines suggest a minimal gradient of 0.2% (0.5% is better) which is good enough to prevent aquaplaning. Another solution is to use vertical alignment deviation tolerance - q to + q. It is, however, very complicated to design and build and mistakes are more likely to be made (Fig. 3).

During rainfall the water flows across the entire road width in micro wheel tracks following the downward gradient of the vertical alignment. "0" gradients in ramp area cause water to pond momentarily. To improve drainage, the width of the road should be divided into working sections.

The German guidelines describe how repairs should be made to eliminate the ponding of water. The solution was applied on the A8 motorway in 2003. diagram of the solution on the A8.

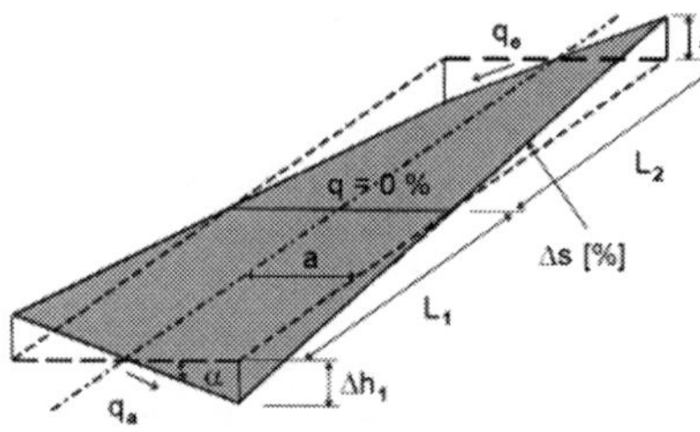

Fig. 3. Diagram of an envelope change of vertical alignment [5].

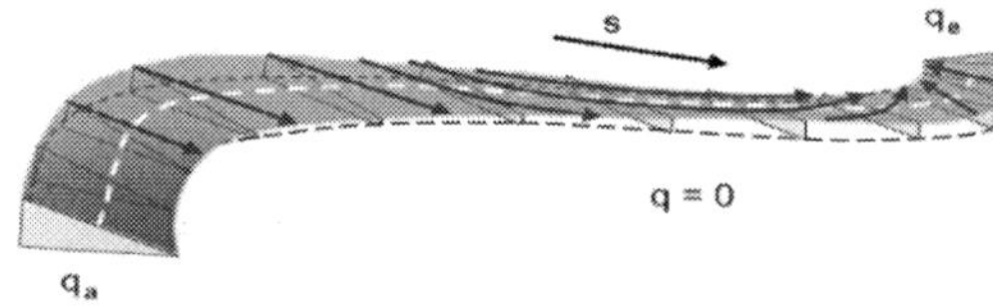

Fig. 4. Diagram of water concentration next to the ramp for "0" change of the vertical alignment, as required in the German guidelines of Bavaria's Ministry [5].

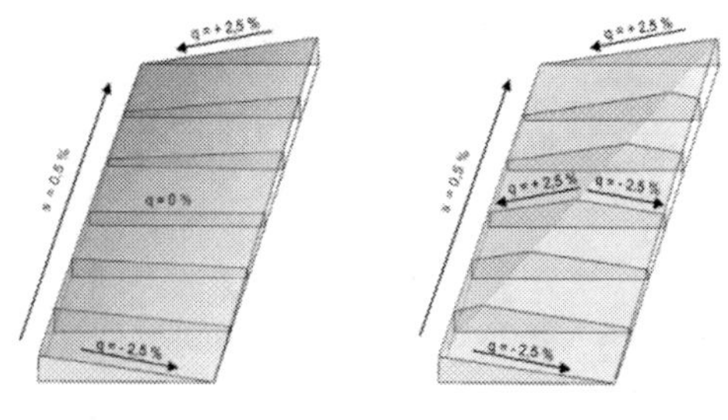

Fig. 5. Example of an envelope applied in 2003 on the A8 motorway.

According to the data of Bavaria's Ministry before the improvement, i.e. between 2000-2002, there were app. 42 accidents recorded in the analyzed area and after the improvement between 2004-2006 there were only 5 accidents related to the problem of water drainage. Figure 5 shows a To ensure good results, the improvement should be designed as follow: move the low gradient vertical alignment to where water can be drained off more easily and increase the approach by increasing the gradient from 0 to 2.5% which will also help to avoid aquaplaning or design an "envelope" relative to height "q".

6 Solution for Poland

In Poland considering the costs of the work, maintenance and additional charges such as road access and use of road charge, the German solution should be modified with the envelope shifted from 1.5% as opposed to Germany's 2.5%. If treated like this, the costs to disassembly the existing road infrastructure can be reduced. This is because of the technical working width of the asphalt spreader (Fig.6) which meant that the existing road infrastructure had to be removed which also included the removal of central reservation or roadside safety barriers near the peak of the envelope.

This solution will partly eliminate aquaplaning and the water level on the road which may cause this phenomenon can be reduced by applying cross slope [9], [10]. For a gradient of 2% the resulting layer will be fairly safe and less than 2 mm. Fig. 7 shows the road geometry before the improvement with blue illustrating the zone of water ponding. The zone is app. 30 running meters which causes aquaplaning for high speeds and heavy rainfall on a long section before the curve.

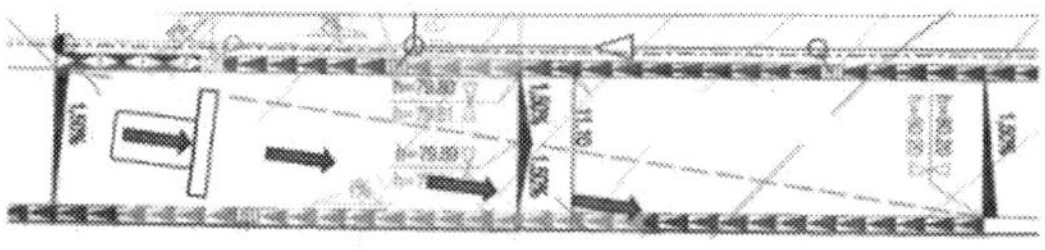

Fig. 6. A section of the design to show the route of the asphalt spreader.

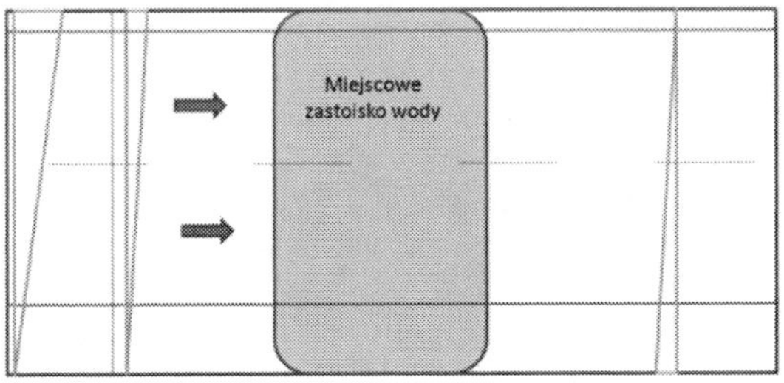

Fig. 7. Condition of the roadway before the repairs with a significant area affected by the ponding of water ("Miejscowe zastoisko wody" (pol.) = the ponding of water).

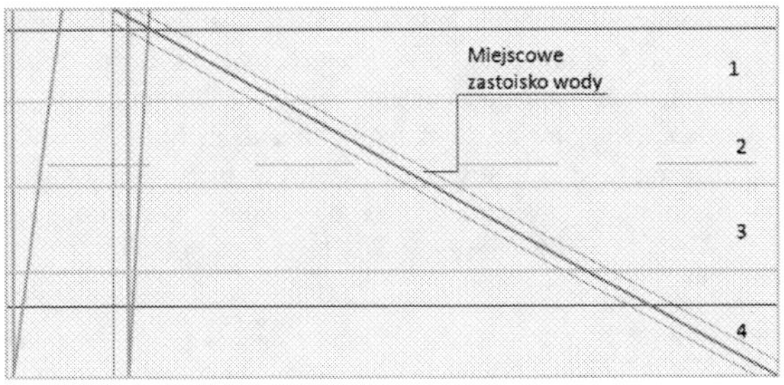

Fig. 8. The approved method for dividing the road works zone into 4 working plots ("Miejscowe zastoisko wody" (pol.) = the ponding of water).

Fig. 8 shows the solution adjusted to the needs of the actual work on the road. The road is divided into 4 working sections to match the technical working width of the asphalt spreader. The resulting ponding of water is now app. 30 cm long as opposed to 30 running meters before the treatment.

The road works were conducted during traffic and the road was closed twice app. 6 days each time. First the emergency lane and 1.5 m of the slow lane were closed. After the work was completed the work was moved to the fast lane and the remaining 2.0 m of the slow lane. Spreading, moving and change of traffic went on for app. 26 hours per each cycle.

7 Conclusion and recommendations

Road accidents at high speed roads characterize specially serious consequences, thus research into road safety is of a special interest [11]. Even if we design and introduce changes to the road gradients correctly, it is uncertain whether unfavorable conditions such as heavy rainfall and excessive speed will not lead to a road accident. Such situations are not uncommon in Poland, and the proof is that road administration of motorways A1 and A4 and regional road 216 has noticed the increased risks within this issue. Thus specially dedicated vertical signs has been introduced in a number of road sections on these roads.

After introducing several years ago the envelope or propeller solutions at some of Polish highway sections, we are sure, supported by German or Swedish experience, that the solution is effective and improves safety. Monitoring of the A1 motorway sections, on which the envelope solution was applied, largely showed total elimination of road accidents during rainfall or immediately after the fall of precipitation. It was also found that the solution is effective only for two lanes roads. In one of the observed places, where there is additionally an ILO inclusion belt, the introduced solution reduced the number of incidents by 62%. At the moment, only vehicles driving at excessively high speed, much higher than allowed, are taking part in this kind of events.

Having this positive experience the next step should be the introduction of regulations. It should go hand in hand with road development and speed increase since thanks to this law any shortcomings in the design and implementation of infrastructure in these critical locations would be avoided. The main challenges related to implementation of these solutions include significant financial expenditures (execution, security of works) and finding a contractor who could undertake the professional implementation of road grades.

References

1. L. Wachnicka, W. Kustra, K. Jamroz, *Development of tools for road infrastructure safety management for the provinces (voivodeships) in Poland*, 26th Conference on European Safety and Reliability (ESREL) (2016)
2. K. Jamroz, L. Michalski, J. Żukowska, *Polish Experience of Implementing Vision Zero*, Current Trauma Reports (2017)
3. Consistency Module Engineers Manual. Interactive Highway Design Model Engineer's Manual. Federal Highways Administration (2012)
4. M. Gaudry, S. Himouri, *DRAG-ALZ-1, a first model on monthly total road demand, accident frequency, severity and victims by category, and of mean speed on highways, Algeria 1970-2007*, Research in Transportation Economics **37**, 1 (2013)
5. K. Jamroz, W. Kustra, M. Budzyński, J. Żukowska, *Pedestrian Protection, Speed Enforcement and Road Network Structure the key Action for Implementing Poland's Vision Zero*, Transportation Research Procedia **14**, pp. 3905-3914 (2016)
6. Råd För Vägars och gators utformning, Trafikverkets Publikation 2012 – Road and street design advice (2012)
7. Oberste Baubehoerde im Bayerischen Staatsministerium des Innern, Verkehrssichere strassengestaltung (2008)
8. M.P. Martinez, A. Garcia, A. T. Moreno, *Traffic Microsimulation Study to Evaluate Freeway Exit Ramps Capacity*, Procedia - Social and Behavioral Sciences **16** (2011)
9. R. Ossowski, K. Gwizdała, *Mechanical Properties of a Dike Formed from a Soil-ash Composite*, Procedia Engineering, **172** (2017)
10. J. Szulwic, P. Tysiąc, *Searching for road deformations using mobile laser scanning*, MATEC Web of Conferences, **122** (2017)
11. J. Szulwic, P. Ziółkowski, A. Janowski, *Combined Method of Surface Flow Measurement Using Terrestrial Laser Scanning and Synchronous Photogrammetry*, Proceedings - 2017 Baltic Geodetic Congress (Geomatics), BGC Geomatics (2017)
12. R. Bergel-Hayat, J. Żukowska J, *Road Safety Trends at National Level in Europe: A Review of Time-series Analysis Performed during the Period 2000–12*, Transport Reviews **35**, 5 (2015)

Risk Diagnosis and Management with BBN for Civil Engineering Projects during Construction and Operation

Marian W. Kembłowski[1], *Beata* Grzyl[1], *Agata* Siemaszko[1*], *Adam* Kristowski[1]

[1]Faculty of Civil and Environmental Engineering, Gdansk University of Technology, Gdańsk, Poland

Abstract. The authors demonstrate how expert knowledge about the construction and operation phases combined with monitoring data can be utilized for the diagnosis and management of risks typical to large civil engineering projects. The methodology chosen for estimating the probabilities of risk elements is known as Bayesian Belief Networks (BBN). Using a BBN model one can keep on updating the risk event probabilities as the new evidence (monitoring information) becomes available. Furthermore, the updated probabilities estimated using the available data for the construction phase serve as background information for the subsequent phase. The integrated two-object model of construction-operation may be then used to optimize the decision making, thus minimizing the risks. To better show how the proposed approach works the authors use the example of the road tunnel constructed and operated under the Dead Vistula River in Gdansk.

1 Introduction

The authors demonstrate how expert knowledge about the construction and operation phases (CP and OP) combined with monitoring data can be utilized for the diagnosis and management of risks typical to large civil engineering projects [1, 2]. The methodology selected for estimating the probabilities of risk events is known as Bayesian Belief Networks (BBN). Using a project-specific BBN model one can keep on updating the risk event probabilities as new evidence (monitoring information/observations) becomes available. Furthermore, the probabilities estimated during the final period of the construction phase with the help the then-available observations serve as background (prior) information for the subsequent (operation) phase (in BBN lingo: the next object). The integrated two-object, construction-operation, model may be used to optimize the decision making process, thus minimizing project risks. To better show how the proposed approach works the authors use the example of the road tunnel constructed and operated under the Dead Vistula River in Gdansk, utilizing the data from tunnel construction subsidence measurements. The importance of the monitoring data collected during the construction activities to risk management during the project operation is demonstrated. We also show how this information affects the estimates of operational risk event probabilities. Some of the risk events that ought to be considered are the occurrence of leaks during the tunnel construction and (improper) subsidence of the tunnel. Such problems may subsequently lead during the tunnel operation to weakening of the tunnel body, and then to growing cracks in the structure. The symptoms of this potential problem can be monitored during tunnel operation.

2 The project description

The project "Connection of Gdańsk Airport with the Port of Gdańsk. The Słowacki Route. Task IV" is part of a broader one (the overall route length is approx. 10 km). The enterprise covers four project tasks of construction and reconstruction of subsequent sections of Słowacki Route, located between the Lech Walesa Airport of Gdansk and the Ku Ujściu traffic interchange.

The major aims of the project are efficiency rise of the Airport - Port of Gdańsk connection, traffic improvement and city development. The Dead Vistula River passage is the project component, implemented by a highway tunnel. Drilling method is applied here on the 1072,5 m long section, by means of the TBM (Tunnel Boring Machine), which includes a 12.6-metre diameter shield.

The Task IV concerning a 2.4 km section covered (see Fig. 1, Fig. 2):

- Słowacki Route construction on the section: Marynarki Polskiej Str. - Ku Ujściu interchange.
- Erecting a 2159 m long engineering structure, including the Dead Vistula River Tunnel (Fig. 3) 1377.5 m long, consisting of two parallel tunnels for each traffic direction separately, including two lanes of traffic.
- Building the Marynarki Polskiej interchange, covering a large roundabout (traffic circle) of a 130 m external diameter, complemented by 7 engineering objects (two road viaducts, a tramway viaduct, two technical viaducts

* Corresponding author: agata.siemaszko@pg.edu.pl

supporting the underground installations along the Marynarki Polskiej St. and two objects for the pedestrians and cyclists).

Fig. 1. Słowacki Route construction - the Task IV, source: http://www.gik.gda.pl/34/inwestycje/budowa_trasy_slowackiego/zadanie_iv.html, access date: 28.06.2015.

Fig. 2. Słowacki Route construction - the Task IV, source: http://www.gik.gda.pl/34/inwestycje/budowa_trasy_slowackiego/zadanie_iv.html, access date: 28.06.2015.

Fig. 3. Słowacki Route construction - the Task IV, source: http://www.gik.gda.pl/34/inwestycje/budowa_trasy_slowackiego/zadanie_iv.html, access date: 28.06.2015.

The ordering party was Gdańsk Commune, represented by Gdańskie Inwestycje Komunalne Sp. z o.o. (GIK Sp. z o.o.). Europrojekt Gdańsk S.A. and SSF Ingenieure GmbH were designers. The contractor was a Spanish company Obrascon Huarte Latin S.A. (OHL).

The tunnel construction is a specified and individual task, involving a number of significant threats. As a result of insufficient background knowledge on the threats connected with tunnel works and their prevention the British Tunnel Society (BTS) published a document "Joint Code of Practice for Risk Management of Tunnel Works in the UK". The code was a joint work of the BTS and British insurance companies. The discussions between the International Tunnel Association (ITA) and the International Association of Engineering Insurers (IMIA) made the code modification and adaptation to international standards. The Code was published again in January 2006 by International Tunnel Insurance Group (ITIG). Its major aim is traffic incident reduction by means of a proper risk identification and effective reduction of risk level at the design and work execution stages of tunnels, caves, shafts and underground connections, complemented by renovation of existing underground infrastructure. The Code is a valuable information source now on the best practice in risk management in the related field.

The Bayesian networks are available tools to analyze the risk linked with design, work execution and operation of building projects. The authors suggest to incorporate Bayesian networks for the risk assessment of a investment project - a giant, technologically demanding, highly cost consuming and unique in a country scale - The highway Dead Vistula River Tunnel.

3 Application of BBNs to a tunnel construction and operation project

We are concerned with risk analysis and management for the phases of construction and operation of a tunnel [3]. Specifically, we are interested in integrating the management of the two phases [4]. Our management model for the construction phase was described in [5]. We assume that the risk analysis information is passed from the construction phase to the operation phase. This is done by estimating for the construction phase the probabilities of two variables:

- tunnel (improper) subsidence,
- faulty element sealing (see Fig. 4).

This information affects in turn the probabilities of two major decision variables for the operation phase, namely: Tunnel Damage, and Tunnel Leaks. These probabilities could be also changed by downward propagation of the information (when available) from: Soil Testing, and tunnel Tightness Monitoring. Finally, these probabilities may be updated by back propagation of the evidence regarding two symptoms: Tunnel Cracks, and Moisture of Tunnel Surface.

https://doi.org/10.1051/e3sconf/20186300004

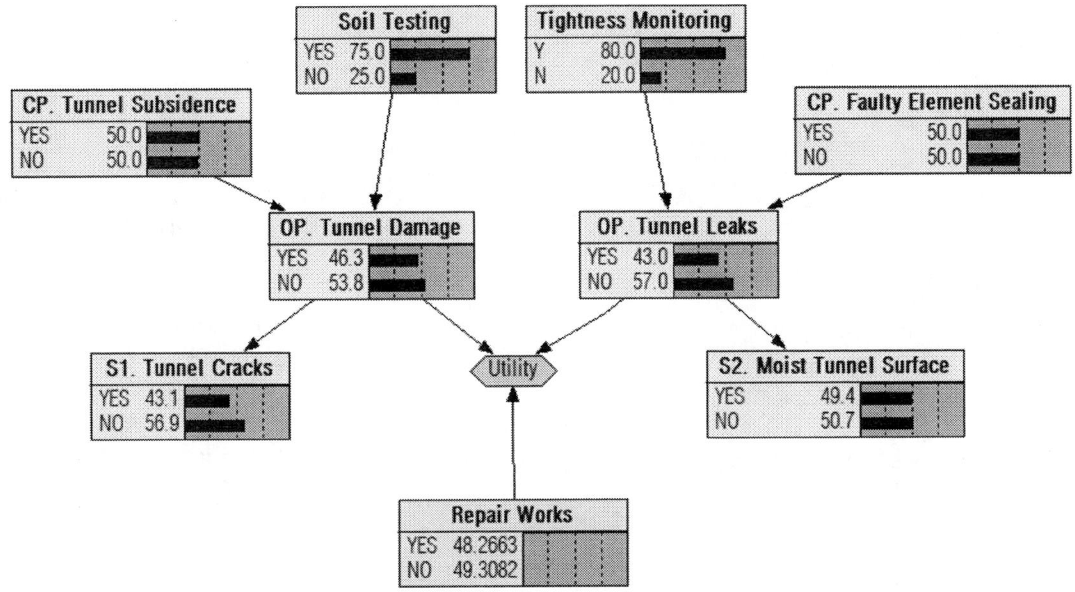

Fig. 4. Bayesian Net for the Tunnel Example.

Fig. 4 shows the connections between the nodes/variables. The strength of the connections is coded in the Bayesian Network by conditional probabilities [5, 6]. In our example the Bayesian Diagnostic/Decision Net Model is used to decide whether to undertake Tunnel Repair Works [10]. In order to do that we define the Utility table, i.e., the utility of each decision for all the combinations of decision variables.

To demonstrate the usefulness of a BBN model to decision making, and thus risk analysis and management, we briefly present and discuss some simulation results [11] obtained using NETICA application. For simplicity we selected as our base scenario (see Fig. 4) the situation where from the construction phase we obtain the information about the two principal variables that is characterized by high entropy. Specifically, we assume 50/50 probabilities of the states (Yes and No) of: Tunnel (improper) Subsidence and Faulty Element Sealing. We also assume that initially we have no new evidence from monitoring the symptoms (Tunnel Cracks, and Moisture of Tunnel Surface). Then we proceed to estimate the expected utilities for different scenarios by changing the information passed from construction phase and introducing new evidence from symptom monitoring.

The first scenario is presented in detail in Fig. 5. It concerns the risk of tunnel settlement due to draining by a TMB machine (Tunnel Subsidence – YES 100%). In the operational stage cracks are observed making it difficult or even impossible to explore the tunnel (Tunnel Craks – YES 100%). These means may lead to the estimation of a decision-change value. Consequently the repair works should start (the YES decision of a 68.42 value).

In the second scenario shown in Fig. 6 the risk is also assumed of tunnel settlement during construction, due to drilling by means of a TMB machine (Tunnel Subsidence – YES 100%). However, the survey during operation detects neither scratches nor cracks in the tunnel structure

(Tunnel Subsidence – NO 100%). This case means the decision change to YES (YES – 49.85). The outlined procedure is a decision-supporting means provided by a tunnel manager. It was correctly assumed that the operation stage is backed up by structural measurements and monitoring.

The measurement discrepancies from the operational ones sign the anomalies, next a decision is made either to start or not the repair activities in the tunnel. The authors' inquiries focus on the tunnel damages and breaking structural tightness. These cases are highly dangerous due to safety of the structure and its users. The damages in the operational stage are triggered e.g. by erecting conditions. In order to avoid such events verification and accounted update of the achieved parameters on the work execution stage of a tunnel construction. This concerns the designed soil-water parameters, next updated due to real conditions affecting tunnel operation, correction of the decision-making parameters, with a continuous update given actual, up-to-date information.

The simulation results are presented in Table 1. Thus observation is possible how the decision value evolved due to risk assessment of unexpected events during the tunnel construction and comparing it with observations of tunnel operation.

A brief analysis of the results shows the relative importance of two issues:

• Information about the tunnel state passed from the construction phase to the operation phase (in here: Tunnel Subsidence and Faulty Element Sealing),

* Corresponding author: agata.siemaszko@pg.edu.pl

• Observing of the tunnel behavior (symptoms) during the operation phase (in here: monitoring Tunnel Cracks, and Moisture of Tunnel Surface).

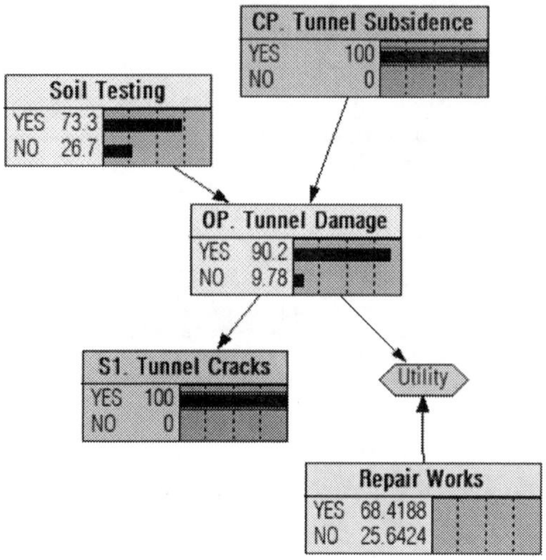

Fig. 5. Network simulation example No. 1

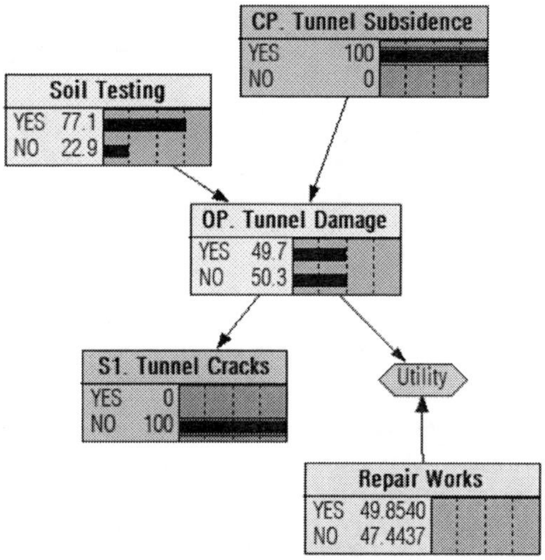

Fig. 6. Network simulation example No. 2.

It is clear that our decision depends strongly on this information. Furthermore, a wrong decision that does not take into account all sources of information may lead to considerable losses. For example: for high entropy information from the construction phase and no symptom information from the tunnel operation phase, Row 1 in Table I, we ought to choose the no Works decision with utility U= 49.31. However, if in reality we have faulty element sealing during the construction phase plus observe leaks during operation (the scenario highlighted in Table 1), and make the same decision not taking account the new evidence, we end up with the utility of U= 27.87. This is a significant utility loss compared to the optimal value (for Repair Works= Yes) of expected utility for this case, namely U= 66.23. The difference may be understood as the risk of making wrong decisions due to the fact of not taking rigorously into account the evidence. A BBN diagnostic/decision tool allows us to avoid this type of risks by taking into account all available evidence and combining it with the expert knowledge (model). It also enables us to estimate expected value of monitoring data - pre a posteriori analysis [7, 8, 9].

Table 1. Summary of simulation scenarios and results.

Tunnel Subsidence (CONSTR PHASE)	**Faulty Element Sealing (CONST RPHASE)**	**Surface cracks (OPER PHASE)**	**Moist surface/ Leaks (OPER PHASE)**	**Expected Utility of Repair Works Decision**	
				Yes	*No*
50/50	50/50	-	-	48.26	49.31
Yes	**50/50**	**Yes**	-	**68.4188**	**25.6424**
Yes	**50/50**	**No**	-	**49.8540**	**47.4437**
50/50	Yes	-	Yes	66.2323	27.8741
50/50	No	-	No	32.8424	67.7095
50/50	No	-	Yes	51.5269	45.4752

4 Summary

The major advantage of the proposed methodology lies in the fact that a created construction-operation risk model works as an integrating tool in which both past and current information about the project and the surrounding conditions is used to assess the risk problems' probabilities at any time of interest. The selection of optimal decisions, based on the assessed probabilities of risk events [12, 13], may be performed, from the viewpoints of the contractor and/or investor. The application of such a Bayesian network model allows one to consider/evaluate decision risks in the environment characterized by considerable uncertainty. The model combines the expert knowledge of the risk problem elements and the cause-effect relationships among them (whose strength is expressed by conditional probabilities) with new evidence (monitoring observations). The evidence from such observations is propagated throughout the whole network resulting in updated (a posteriori) probabilities of the events/variables represented in the model. The updated risk probabilities that reflect the cumulative effect of all new evidence may lead us in turn to new (updated) selection of optimal decisions.

It is worthwhile to mention that the model variables and the relationships among them are represented by easy to understand, discuss, and modify graphs. This characteristic is of importance during the process of creating and testing the model by a group of problem experts and knowledge engineers.

Furthermore, as our knowledge regarding new factors influencing risk issues becomes broader we may readily incorporate the new relationships in our integral risk model. This is due to the elasticity of Bayesian nets, namely their ability to accept new interconnected events with multiple connections to the existing one.

References

1. Polska Norma PN-EN 60300-3-3:2006 „Część 3-3: Przewodnik zastosowań. Szacowanie kosztu cyklu życia" (2006) (in polish)
2. M. Apollo and E. Urbańska-Galewska, *Model of Risk Assessment (MORA) concept for the investment part of urban regeneration projects*, Environmental Engineering, Proceedings of the International Conference on Environmental Engineering. ICEE. **9**. (2014)
3. B. Grzyl and A. Kristowski, *BIM jako narzędzie wspomagające zarządzanie ryzykiem przedsięwzięcia inwestycyjnego*, Materiały Budowlane, **6**, pp. 52-54, (2016) DOI: 10.15199/33.2016.06.22
4. E. Miszewska-Urbańska, *Modern Management Challenges of Floating Housing Development*, Real Estate Management and Valuation, **24**, issue 1, pp. 31-40, (2016)
5. M. W. Kembłowski, B. Grzyl, A. Kristowski, and A. Siemaszko, *Risk Modelling with Bayesian Networks - Case Study: Construction of Tunnel under the Dead Vistula River in Gdansk*, Procedia Engineering **196**, 585-591, (2017)
6. U. B. Kjaerulff and A. L. Madsen, *Bayesian Networks and Influence Diagrams. A Guide to Construction and Analysis*, Springer Science+Business Media, LLC, (2008)
7. Norsys. [Online] Available: http://www.norsys.com/netica. [Accessed: 09.2017]
8. M. Miskiewicz, L. Pyrzowski, and K. Wilde, *Structural Health Monitoring System for Suspension Footbridge*, in Proc. of 2017 Baltic Geodetic Congress (BGC Geomatics), pp. 321-325, (2017) DOI: 10.1109/BGC.Geomatics.2017.8
9. M. Apollo, E Miszewska-Urbańska, *Analysis of the increase of construction costs in urban regeneration projects*, Advances in Science and Technology Research Journal, **9.28**, (2015)
10. M. Kurpińska and A. Małasiewicz, "Der Betonschutz vor der Wirkung der aggressiven Umbebung," Ibausil 15.Internationale Baustofftagung 24 -27 September 2003, F.A. Finger – Institut für Baustoffkunde, Bauhaus-Universität Weimar, Weimar, Bundesrepublik Deutschland, 2-0943 – 2-0952, (2003)
11. D. Fudenberg and J. Tirole, *Game Theory*, MIT Press, (1991)
12. J. Levin, *Dynamic Games with Incomplete Information,* February (2002), http://web.stanford.edu/~jdlevin/Econ%20203/DynamicGames.pdf
13. N. Fenton, M. Neil, *Risk Assessment and Decision Analysis with Bayesian Networks*, CRC Press, ISBN: 9781439809105, (2012)

Effect of GIS parameters on modelling runoff from river basin. The case study of catchment in the Puck District

Dominika Kalinowska[1,*], *Paweł* Wielgat[1], *Tomasz* Kolerski[1], and *Piotr* Zima[1]

[1] Department of Hydraulic Engineering, Gdansk University of Technology, Gdansk, Poland

Abstract. The study was preceded to check the impact of the adopted digital elevation model resolution on the determination of catchment parameters and the further influence of the received information on calculations related to the hydrograph at the closing point of the basin after a synthetic 100-year precipitation episode. The study area are river basins located in the Puck commune convoying water directly to the Puck Bay. These studies allowed the formulation of the basic assumptions for the research project (in short called WaterPuck) supported by the National Centre for Research and Development.

1 Introduction

The catchment's response to atmospheric precipitation is determined by many factors, including landform [1, 2]. The basin delimitation depend on the direction of exposure of the sloping lands (aspect) and orientation of analyzed area and lead to basic geographic unit, a watershed, which is the limited space from which all the water is collected and drains surface water to a common point - the receiver (e.g. a river, a reservoir). The size of the basin has a direct impact on the volume of collected water, which need to be managed according to the local requirements. It seems difficult or even impossible to determine experimentally in which direction rainwater flows from each part of the analyzed area. Catchment boundaries are estimated based on topographic maps or a digital elevation model (DEM). In both cases, the results obtained depend on the accuracy of the input material and the simplifications adopted by the person or program processing the topographic information. Accuracy problems may occur while maps with a large scale or too low resolution of the digital elevation model is analyzed. It could be additionally hampered by employing an approximate approach to delineating the catchment boundaries.

The outflow from the basin is unsteady due to number of external inputs which sometimes may cause the flow to be characterized by rapid variability in time. From the flood hazard point of view, the most dangerous situation is overlapping flood surges originating from several subbasins in time and space [3]. The shape of the surface runoff hydrograph is determined, among others, by concentration time. It is a time required to describe the outflow delay from individual sub-basins. By definition this is the time after which the water from the most hydraulically advantageous point will flow down to the point closing the catchment. The calculation formulas use the lengths of the surface runoff lines and slopes for individual subbasins - these are further parameters resulting directly from the assumed altitude elevation mapping.

In the era of digitization and the multitude of available data, a large part of the work is improved by GIS tools [4–8]. In this paper, the calculation were made by adopting Soil and Water Assessment Tool (SWAT) software [9–11] used as a plug-in for the Q-GIS program (QSWAT). QSWAT uses Terrain Analysis Using Digital Elevation Models (TauDEMs) for watershed delineation. TauDEM provides a suite of programs to perform various geoprocessing functions [12–14]. In the first steps, the user defines how many cells of the numerical terrain model per square kilometer will be used for further pro-cesses. The preliminary delineation of the catchment boundaries is based on local linear hollows accepted as a river net-work [15]. The higher density of DEM cells included, the more streams are visible, and thus the more subbasins will be designated (each stream is assigned its own subbasins). The adoption of other subbasin borders will also affect local slope be-cause the slope determined for each calculation cell is ultimately averaged as a slope corresponding to the dedicated subbasin. The use of GIS tools allows for a large detailing and fragmentation of the hydrographic division, which will affect the outflow model from the basin.

The study was preceded to check the impact of the adopted DEM resolution on the determination of catchment parameters and the further influence of the received information on calculations related to the flow rate at the closing point of the basin after a synthetic 100-year precipitation episode (probability of exceeding p = 1%). Calculations were made for two digital elevation models: (1) CODGiK (Polish Central Centre for Geodetic and Cartographic Documentation) and (2) DEM provided by the European Environment Agency. The study area are river basins located in the Puck commune convoying water directly to the Puck Bay.

* Corresponding author: domszawu@pg.edu.pl

https://doi.org/10.1051/e3sconf/20186300005

These studies allowed the formulation of the basic assumptions for the research project (in short called WaterPuck) supported by the National Centre for Research and Development within the BIOSTRATEG III program No. BIOSTRATEG3/343927/3/ NCBR/2017. Under this program, a system to calculate the maximum allowable amount of fertilizers to be used on fields, together with the determination of their impact on the environment of the Baltic Sea in the area of the Puck Bay will be created.

GIS technology and data analysis is very useful for hydro-logical modelling and determination of the uncontrolled basin outflow. Nowadays, through developed surveying tools including laser scanning (LIDAR) or photogrammetric image processing high accuracy digital elevation model can be generate easily [16–19]. For large watersheds though is still hampered by computing power. In the paper hydrologic rain-fall-runoff model was implemented to catchments in Puck region by using three DEM models with resolution 50x50 m, 10x10 m and 1x1 m. In addition the threshold parameter of minimum basin area was tested. Hydrologic model was fed with synthetic rainfall and real storm observed in July 2018.

2 Area of study

The study are concerns Puck Bay, where surface water flows from agricultural areas of the Puck District (Poland, Pomeranian Voivodship) (Fig. 1). This area is part of the Kashubian Coastland and includes Kepa Pucka together with two ice-marginal valleys: Plutnica Valley and Kashubian Valley. It is a young glacial area with a varied sculpture characterized by high denivelation (from -0.5 to 113.5 m above sea level) and cut by smaller valleys with steep slopes [20]. The bottoms of the valleys are river beds (e.g. Gizdebka River, Bladzikowski Creek). The geological structure consists of fluvioglacial deposits, mainly sands and gravels interlaced with clays. In terms of land use, we distinguish mainly two forms: agricultural and forest areas.

Fig. 1. Map of the study area (source: Google Maps, accessed 10 June 2018).

3 Methodology and materials

Data form laser scanning were obtained from Central Documentation Centre of Geodesy and Cartography in War-saw. The data were used to generating DEM with variable resections (1, 5 and 10 m raster) – Fig. 3. Basin areas, rivers lengths and terrain denivelation were automatically obtained by using SWAT model. In calculations the threshold parameter for minimum basin area was tested using 1, 5 and 10 km^2 (Fig. 4).

Soil types and land use was determined on the base of soil maps obtained from Marshal's Office of Pomeranian Voivodship. The maps were used to establish the soil groups based on the surface soils. Land use was also determined on the soil maps however to make the analysis less dependent on this parameter, only four types of land development was used in the study, i.e. forestland, meadow, cultivated agricultural land (row crops), residential (average lot size ¼ acre) and fully developed urban areas. All above information were used to determine runoff curve numbers (CN) for average watershed condition [21, 22].

Hydrological calculations were proceeded for all combinations of DEM accuracies and minimum basin area. Model input was the 1% synthetic precipitation with duration of 60 minutes. Various methods for determining synthetic rain are used [23]. The rainfall time distribution was specified according to DVWK formula as shown in Fig. 2. Real precipitation measured in night from July 17th to 18th 2018 was also used to compare model results with real data. Water surface elevation is surveyed on gauging stations located on each river's outlet to the Puck Bay. The discharge is calculated from the rating curve established for each station.

Hydrological modelling System HEC HMS was used to in this study to evaluate rainfall-runoff modelling [24, 25]. The software was developed by U.S. Army Corps of Engineers and provided on the Corps' web page as an open access engineering tool. All required data were generated by using used as a plug-in for the Q-GIS program.

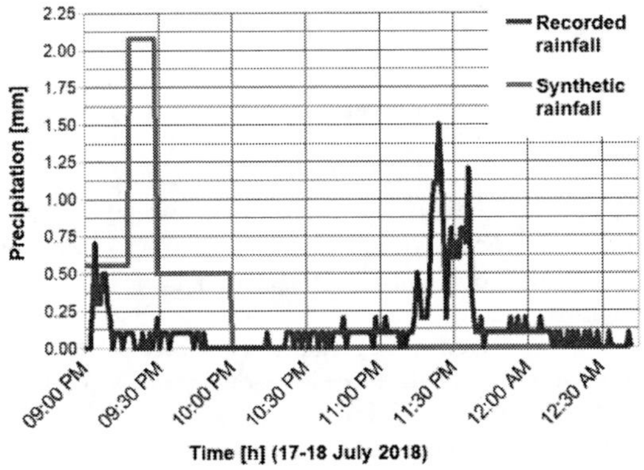

Fig. 2. Synthetic and recorded rainfall used in simulations.

 https://doi.org/10.1051/e3sconf/20186300005

It is worth emphasizing the long time needed to generate a high resolution DEM grid. In the case of the tested catchment area, it was from 2 hours for 50x50 m resolution to about 72 hours of continuous virtual machine operation for 1x1m resolution. For DEM with the lowest resolution, basins were generated in no more than half an hour, while for the DEM with a resolution of 1x1m it was about 36 hours. During the generation of these grids, the resources of the PIONIER cloud network service with 3 virtual machines (6 CPUs, 8 GB RAM) were used.

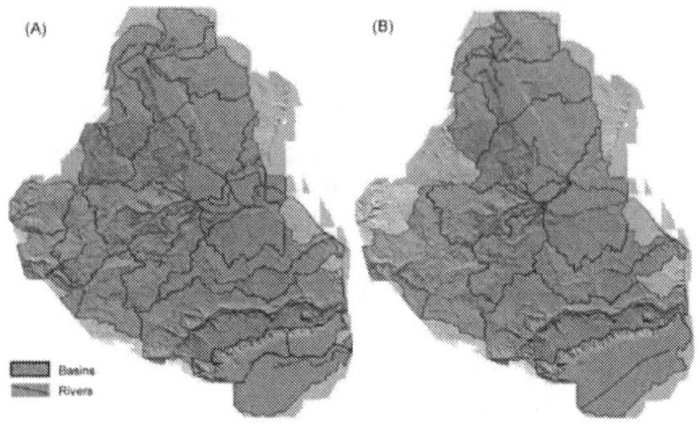

Fig. 3. Subasins distribution In Puck district for using DEM with (A) high accuracy (1x1 m) and (B) low accuracy (10x10 m) (source: SWAT).

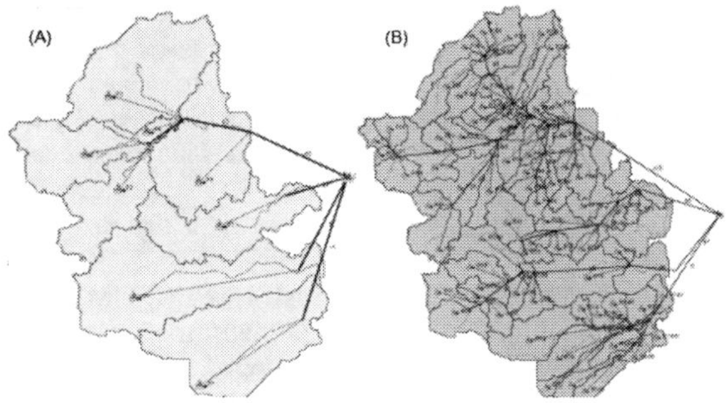

Fig. 4. Basin model and river network in HEC HMS model for low (A) and high (B) accuracy catchments (source: HEC-HMS).

4 Results and discussion

As a result of hydrological simulations, a total outflow was obtained in the characteristic cross-sections of the prepared catchment models. These values for each scenario are similar. The results in the form of maximum flow rate (peak dis-charge), time of its occurrence and total volume of water are presented in Tables I – IV and Figures 5 – 6.

Differences in the results obtained are related to the level of details included in individual models. When the DEM resolution is low many elements, i.e. road embankments and levees are invisible. In the case of high DEM resolution, very ac-curate DEM takes into account these details and fundamentally changes the sub-basins delimitation along the crown of linear objects (Fig. 7).

Table 1. Discharge to Bay of Puck synthetic rainfall.

Probability 1%, duration 60 minutes					
Resolution of DEM [m²]	1	100	100	100	2500
Minimal area subbasin [km²]	5	1	5	10	5
Time of peak discharge [HH:MM]	4:29	3:23	4:41	11:46	3:33
Peak discharge [m³/s]	83,7	84,5	61,5	22,9	97,7
Total volume [1000 m³]	1363,1	1343,0	1303,6	1301,6	1272,8

Table 2. Discharge to Bay of Puck recorded rainfall.

Recorded rainfall 7/17/2018					
Resolution of DEM [m²]	1	100	100	100	2500
Minimal area subbasin [km²]	5	1	5	10	5
Time of peak discharge [HH:MM]	6:36	5:36	6:26	14:21	5:47
Peak discharge [m³/s]	8,0	7,8	6,5	2,2	9,9
Total volume [1000 m³]	138,6	142,1	131,4	131	130,4

Table 3. Gizdebka river discharge recorded rainfall.

Recorded rainfall 7/17/2018					
Resolution of DEM [m²]	1	100	100	100	2500
Minimal area subbasin [km²]	5	1	5	10	5
Time of peak discharge [HH:MM]	6:21	7:46	7:58	20:35	6:25
Peak discharge [m³/s]	2,2	3,3	1,5	0,4	2,0
Total volume [1000 m³]	37,1	35,9	36	34,7	33,4

Table 4. Gizdebka river discharge synthetic rainfall.

Probability 1%, duration 60 minutes					
Resolution of DEM [m²]	1	100	100	100	2500
Minimal area subbasin [km²]	5	1	5	10	5
Time of peak discharge [HH:MM]	4:06	7:33	5:42	18:24	4:09
Peak discharge [m³/s]	23,6	36,0	15,7	4,5	20,9
Total volume [1000 m³]	389,4	376,7	377,2	362,5	350,3

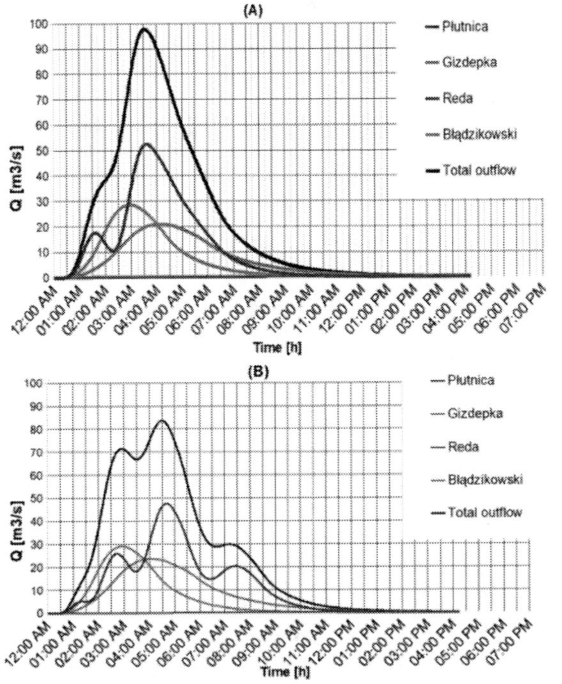

Fig. 5. Simulated hydrographs from Puck District (for precipitation with a probability of exceeding p = 1 % and rain duration 60 min) on the outflow to the Puck Bay for variable digital terrain model: (A) 50 m2 and (B) 1 m2.

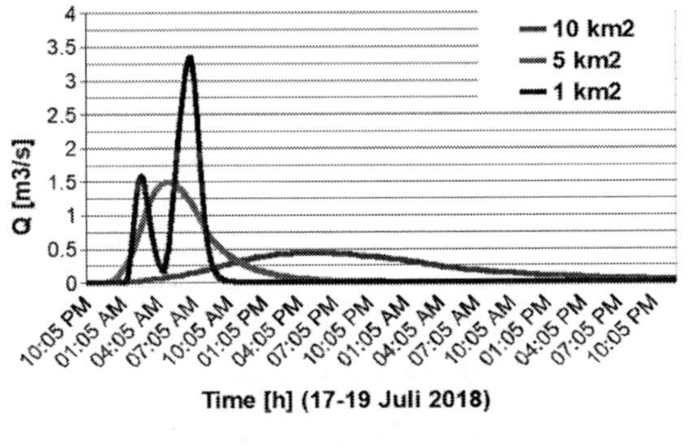

Fig. 6. Simulated water discharge at the outlet of the Gizdebka River to the Puck Bay, for rainfall recorded on June 17th – 18th, 2018.

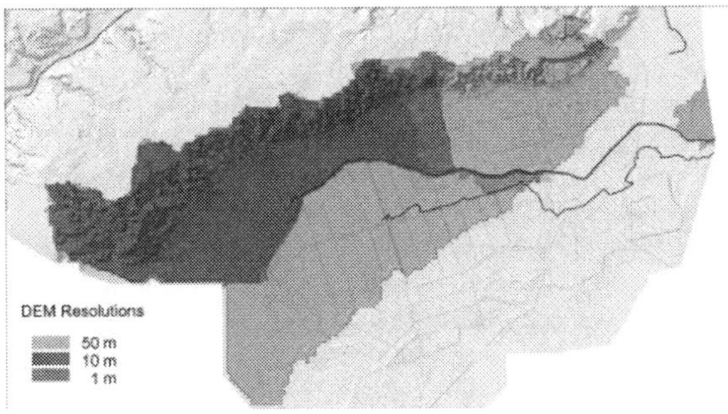

Fig. 7. Differences in Reda catchment for variable digital terrain model – 1, 10 and 50 m^2.

5 Summary and conclusion

Base on the study following conclusions could be stated:

1) Resolution of the digital terrain model must be properly chosen to avoid following problems:
a) too high resolution of the DEM model i.e. very detail terrain reconstruction will false watershed divides. This is usually caused by linear structures: berms, irrigation channels, embankment dikes and others. This will cause not realistic lag time for the sub-basins.
b) too low resolution (50x50 m) will affect missing small streams and irrigation channels and in consequence the river network will be not be completed. Sub-basins are coarse with large area not reflecting the reality.
c) the DEM resolution affects the basin shape in the downstream reach, close to the outlet. It is caused by the direct sea basins which areas changes with model resolutions.

2) The threshold parameter of minimum basin size has significant impact on the outflow formation form the watershed. The volume of the surface outflow similar, but the peak discharge varies significantly. Also the time to peak value is different for variable minimum basins parameters, in such a way that the closed to reality is division to average sub-basins (5 km^2). In that case the peak discharge calculated in the Gizdebka River was in rage of 1.5 m^3/s , very close to the measured water outflow from the survey.

Online references will be linked to their original source, only if possible. To enable this linking extra care should be taken when preparing reference lists.

Acknowledgment

The authors are grateful to CI TASK (Centre of Informatics - Tricity Academic Supercomputer & network) for granting access to the computing infrastructure built in the projects No. POIG.02.03.00-00-028/08 "PLATON - Science Services Platform" and No. POIG.02.03.00-00-110/13 "Deploying high-availability, critical services in Metropolitan Area Networks (MAN-HA)"

These studies allowed the formulation of the basic assumptions for the research project (in short called WaterPuck) supported by the National Centre for Research and Development within the BIOSTRATEG III program No. BIOSTRATEG3/ 343927/3/NCBR/2017. Under this program, a system to calculate the maximum allowable amount of fertilizers to be used on fields, together with the determination of their impact on the environment of the Baltic Sea in the area of the Puck Bay will be created.

References

1. D.R. Maidment, *Handbook of hydrology*, McGraw-Hill (1992)

2. S. Eslamian, *Handbook of engineering hydrology: fundamentals and applications*, CRC Press, Taylor and Francis NY (2014)

3. W. Szpakowski and M. Szydłowski, *Evaluating the Catastrophic Rainfall of 14 July 2016 in the Catchment Basin of the Urbanized Strzyza Stream in Gdańsk*, Polish Journal of Environmental Studies, **27**, 2, pp. 861-869, (2018). DOI: 10.15244/pjoes/75962

4. M. Szydłowski, et al., *Stormwater and snowmelt runoff storage control and flash flood hazard forecasting in the urbanized coastal basin*, in Proc. of 14th International Symposium Water Management and Hydraulic Engineering WMHE 2015, pp. 141-150 (2015)

5. P. Wielgat and P. Zima, *Analysis of the impact of the planned sewage discharge from the 'North' Power Plant on the Vistula water quality*, in Proc. of 16th International Multidisciplinary Scientific GeoConference SGEM 2016, book 3, vol. 3, pp. 19-26 (2016). DOI: 10.5593/SGEM2016/HB33/S02.003

6. P. Zima, *Modeling of the Two-Dimensional Flow Caused by Sea Conditions and Wind Stresses on the Example of Dead Vistula*, Polish Maritime Research, **25**, s1, pp. 166-171 (2018). DOI: 10.247/pomr-2018-0038

7. T. Kolerski, *Modeling of ice phenomena in the mouth of the Vistula River*, Acta Geophysica, **62**, 893–914 (2014). DOI:10.2478/s11600-014-0213-x

8. T. Kolerski, *Ice cover progression due to flow regulation at the Wloclawek dam*, Acta Scientiarum Polonorum. Formatio Circumiectus, **14**, 229–240 (2015)

9. J.G. Arnold, N. Fohrer, *SWAT2000: Current capabilities and research opportunities in applied watershed modelling*, Hydrological Processes, **19**, 3, pp. 563–572, (2005). DOI: 10.1002/hyp.5611

10. J.G. Arnold, et al., *Soil and Water Assessment Tool input/output file documentation: Version 2012*, Texas Water Resources Institute, TR-439, (2012)

11. R. Srinivasan, et al., *Large area hydrologic modeling and assessment part II: model application*, Journal of the American Water Resources Association, **34** (1), pp. 91-101, (1998). DOI: 10.1111/j.1752-1688.1998.tb05962.x

12. T.K.Tesfa, et al., *Extraction of hydrological proximity measures from DEMs using parallel processing*, Environmental Modelling & Software, **26** (12), pp. 1696-1709, (2011). DOI: 10.1016/j.envsoft.2011. 07.018

13. J.P. Wilson, *Digital terrain modeling*, Geomorphology, **137**, 1, pp. 107-121 (2012). DOI: 10.1016/j.geomorph.2011.03.012

14. Y.T. Dile, et al., *Introducing a new open source GIS user interface for the SWAT model*, Environmental Modelling & Software, **85**, pp. 129-138 (2016). DOI: 10.1016/j.envsoft.2016.08.004

15. D.G. Tarboton, *A new method for the determination of flow directions and upslope areas in grid digital elevation models*, Water Resources Research, **33**, 2, pp. 309-319 (1997). DOI: 10.1029/ 96WR03137

16. K. Bobkowska, A. Inglot, M. Mikusova, P. Tysiąc, *Implementation of Spatial Information for Monitoring and Analysis of the Area Around the Port Using Laser Scanning Techniques*, Polish Maritime Research, **24**, S1, pp. 10-15 (2017). DOI: 10.1515/pomr-2017-0015

17. K. Bobkowska, A. Janowski, J. Szulwic, *3D modelling of cylindrical-shaped objects from LiDAR data - an assessment based on theoretical modelling and experimental data*, Metrology and Measurement Systems, Vol. 25, issue 1, pp. 47-56, 2018. DOI: 10.24425/118156

A. Inglot, P. Tysiąc, *Airborne Laser Scanning Point Cloud Update by Used of the Terrestrial Laser Scanning and the Low-Level Aerial Photogrammetry*, Proceedings - 2017 Baltic Geodetic Congress (Geomatics), BGC Geomatics 2017, pp. 34-38 (2017). DOI: 10.1109/ BGC.Geomatics.2017.75

18. J. Szulwic, P. Ziolkowski, A. Janowski, *Combined Method of Surface Flow Measurement Using Terrestrial Laser Scanning and Synchronous Photogrammetry*, Proceedings - 2017 Baltic Geodetic Congress (Geomatics), BGC Geomatics 2017, pp. 110-115 (2017). DOI: 10.1109/ BGC.Geomatics.2017.54

19. D. Potrykus, et al., *Assessing groundwater vulnerability to pollution in the Puck region (denudation moraine upland) using vertical seepage*

https://doi.org/10.1051/e3sconf/20186300005

method, E3S Web of Conferences. **44**, 00147 (2018). DOI 10.1051/ e3sconf/20184400147

20. Jr.A.T. Hjelmfelt, *Investigation of curve number procedure*, Journal of Hydraulic Engineering, **117**, pp. 725–737 (1991). DOI: 10.1061/(ASCE)0733-9429(1991)117:6(725)
21. SCS: National Engineering Handbook, Section 4: Hydrology, Soil Conservation Service, USDA, Washington, D.C. (2004)
22. K. Weinerowska-Bords, *Development of Local IDF-formula Using Controlled Random Search Method for Global Optimization*, Acta Geophysica, **63**, 1, pp 232–274, (2015). DOI 10.2478/s11600-014-0242-5
23. W. A. Scharffenberg and M. J. Fleming, *Hydrologic Modeling System HEC – HMS User's Manual* (2010)
24. HEC-HMS Technical Reference Manual, US Army Corps of Engineers Hydrologic Engineering Center, USA (2000)

Load test of new European record holder in span length among extradosed type bridges

Mikołaj Miśkiewicz[1,*], and *Łukasz* Pyrzowski[1]

[1]Dep. of Mechanics of Materials and Structures, Faculty of Civil and Environmental Engineering, Gdańsk University of Technology, G. Narutowicza 11/12, 80-233 Gdańsk, Poland

Abstract. The article presents acceptance load tests of a newly built extradosed MS-3 bridge located along the national road DK-16 near Ostróda (Poland). The structure features significant dimension in European reference scale. It is a new record holder in span length regarding this distinct bridge category. Its length ranges 206 m. Static and dynamic load tests were performed here. The program was extended by entire structural laser scanning and identification of natural vibration forms.

1 Introduction

An extradosed bridge is a structure combining the technical solutions for a prestressed box girder bridge and a cable-stayed bridge [1]. The extradosed concept precursors are Ganter Bridge in Switzerland and bridge in Rzuchów in Poland, both built in 1980. Nevertheless, Jacques Mathivat is most commonly credited as an inventor of extradosed terminology and its design concepts by publishing his ideas in 1988 [2]. Since that time, this type of a bridge is under current development and new modern, record-breaking structures are permanently erected. Currently the world longest extradosed bridge, according to structurae.net, is Arrah-Chhapra Bridge in India with 1920 m of total main bridge length (16 spans, 120 m each), built in 2017. In turn, the longest span world record belongs to Kiso Gawa Bridge in Japan, 275 m long. The European bridges reach lower achievements. However, one of the records belongs to a Polish structure. It started in 2013, when the extradosed bridge was completed in Kwidzyn over Vistula River. This structure, with its main span length of 204 m, became the record holder in this category in Europe [3]. Recently, at the end of 2017, the bridge MS-3 construction was finished along the road DK-16 near Ostróda. The structure was designed by Transprojekt Gdański Sp. z o.o. and erected by Budimex S.A. The longest European span length achievement has been taken over and raised up to 206 m. The paper deals with the acceptance tests of this bridge, undertaken 14th of December 2017. The tests were designed and performed by Bridge Laboratory of Department of Mechanics of Materials and Structures at Gdansk University of Technology and Aspekt Laboratorium Sp. z o.o. Due to the fact of significant structural size in European scale, the tests have been assumed an extended program, related to typical structure, including e.g. structural laser scanning. The load tests were prepared and performed taking into account experience of the paper authors [4-12] as well as other contributors in the field [13-23].

2 Structure description

The MS-3 bridge is located along the national road DK-16, designed to cross the deep valley and the Ornowska Struga stream near Ostróda. It is a continuous four-span structure, in its majority made of prestressed concrete. It includes three short towers (deviators) and a system of stayed cables, to be eventually called an extradosed bridge, see Figure 1.

Fig. 1. Side view of MS-3/DK-16 bridge.

The theoretical total structural length is 677 m, the spans lengths are 132.5 m, 206.0 m, 206.0 m and 132.5 m. The towers rise above the deck to a height of 23 m. They are made of C60/75 reinforced concrete (two pylons) and prestressed concrete (one pylon). The total deck width is 28.6 m, consisting of two roadways, three service sidewalks and four bands. The deck structure is a three-chamber box made of prestressed C60/75 concrete,

* Corresponding author: mmisk@pg.edu.pl

 https://doi.org/10.1051/e3sconf/20186300006

see Figure 2. The height of the box varies between ~4 m in spans up to ~6 m above middle supports. Steel of a characteristic 1860 MPa strength is used for prestressing, while A-IIIN B500SP steel is used for soft reinforcement. The external prestressing on each pylon is carried out with the use of 167L15.7 tendons for the three longest cables, 139L15.7 for two consecutive intermediate cables and 135L15.7 for two shortest cables. The bridge is designed for the highest load class A according to Polish standard [24].

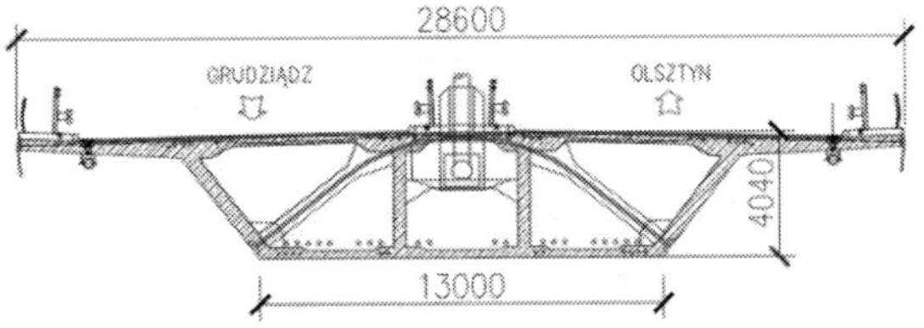

Fig. 2. Typical cross-section of MS-3/DK-16 bridge.

3 Computational analysis

In order to create the acceptance load test program, i.e. ballast choice selection of measurement points location and adequate sensor choice, it was necessary to create a relevant numerical model of the structure. The Finite Element Method (FEM) model (see: [25] and [26]) was made within the SOFiSTiK system, see Figure 3.

The adopted model involves three finite element types, marked with symbols (R), (P) and (K):
(R): 1-dimensional, 2-node, Timoshenko spatial rod-type elements, class C° with linear shape functions, considering shear effect and eccentricity of the beam axis: 1650 elements;
(P): 2-dimensional, 4-node, Timoshenko-Reissner shell-type elements, class C° with bi-linear shape functions and deformation enhancement in the surface including shear and eccentricity: 161289 elements;
(K): 1-dimensional, 2-node, spatial truss-type elements, class C° with linear shape functions: 42 elements.

Fig. 3. Numerical model of MS-3/DK-16 bridge.

In the course of static simulations displacements, internal forces and stresses were determined in individual structural components under the action of standard-based road loads and the test load setting schemes. The results of dynamic calculations were frequencies and modal shapes of natural vibrations.

4 Static tests

In the course of static tests three different load settings were implemented. The ballast was assumed in the form of 5-axle vehicles, each one of a total 442 kN weight. Two settings (U34 and U45) were designed to obtain maximum cross-sectional bending moment in the middle of spans 3-4 and 4-5, respectively (16 vehicles in each test), the third one (U3) to obtain minimum cross-sectional bending moment above P3 support (36 vehicles). The extreme values of the above parameters were established in the range of 75-100% compared to the results obtained for a road load model according to [24]. The load settings U34 and U45 are shown in Figures 4 and 5.

Fig. 4. Full ballast loading during U34 setting.

Fig. 5. Full ballast loading during U45 setting.

In the coverage of load tests, the following measurements were performed:
- vertical displacements of the span by means of precise levelling (Carl Zeiss Ni007), optical tachymeters (LEICA NOVA TM 50) and dial gauges: 17 sections, 2 points per each located at the ends of cantilevers = 34 measurement points,
- three-directional displacements of pylons by an optical tachymeter (LEICA NOVA TM 50): 3 measurement points,
- force in cables by accelerometers using vibration method (APEK, AV32M37R4.6U16): 8 measurement points,

 https://doi.org/10.1051/e3sconf/20186300006

- stress (strain) of the deck by stringed extensometers (GEOKON 4000A-1): 2 measurement points located at bottom plate of the girder box,
- settlement of supports by means of precise levelling (Carl Zeiss Ni007): 8 measurement points.

The localisation of all measurement points due to static load tests is presented in Figure 6A.

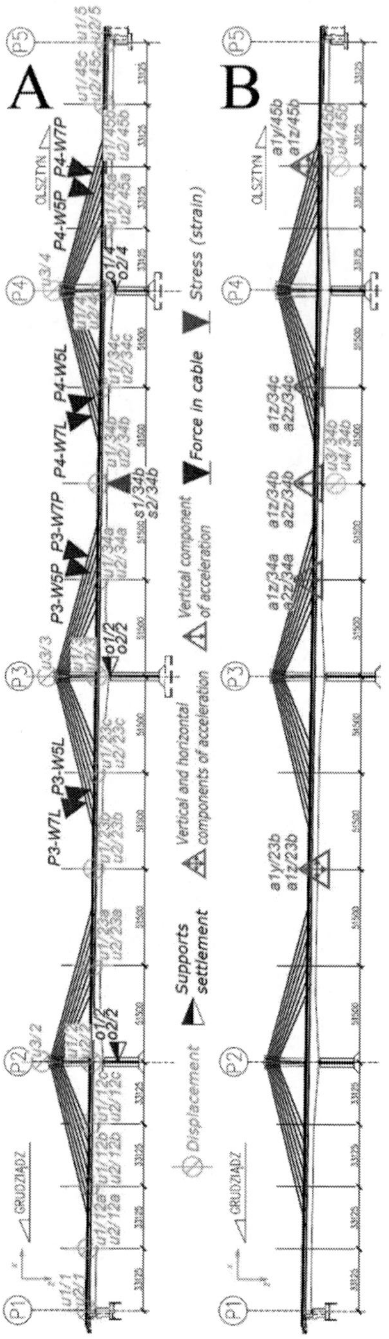

Fig. 6. Measurement point localisation for static load tests: left side (A), dynamic load test: right side (B).

Table 1 presents middle span vertical displacements measured during static settings with reference to the theoretical values.

Table 1. Measured and calculated middle span vertical displacements of deck obtained in static tests.

Ballast setting	Measure point	Elastic displ. Uz [mm]	Theoret. displ. Uz [mm]	Accuracy [%]
U34 full load	u1/23b	-55.5	-85.4	65
	u1/34b	155.0	183.0	85
	u1/45b	-49.5	-66.6	74
U45 full load	u1/34b	-50.4	-64.5	78
	u1/45b	56.5	92.9	61
U3 full load	u1/23b	35.5	55.1	64
	u1/34b	44.7	55.2	81
	u1/45b	-11.8	-18.3	64

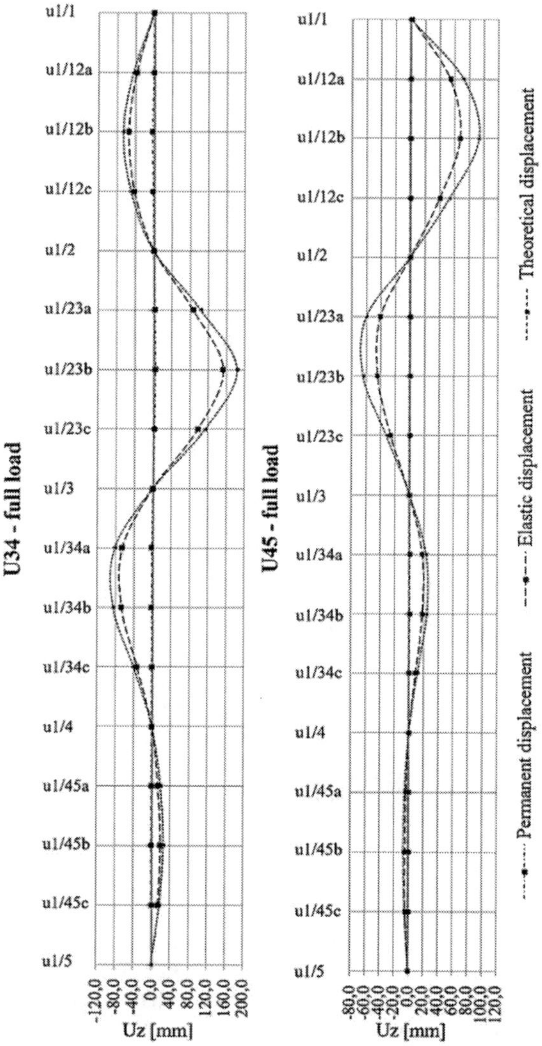

Fig. 7. Measured and calculated displacements along the structure obtained in U34 and U45 static settings.

Figure 7 shows example results of vertical displacements along the bridge due to full U34 and U45 settings, compared with the theoretical values. Table 2 presents horizontal displacements of the pylons and Table 3 force increments in external cables, both referred to the theoretical values.

Table 2. Measured and calculated horizontal displacements of pylons obtained in static tests.

Ballast setting	Measure point	Elastic displ. Ux [mm]	Theoret. displ. Ux [mm]	Accuracy [%]
U34 full load	u3/2	-9.9	-5.7	174
	u3/3	31.0	52.6	59
	u3/4	-31.8	-28.3	112
U45 full load	u3/2	2.7	1.7	159
	u3/3	-8.2	-15.1	54
	u3/4	19.8	27.8	71
U3 full load	u3/2	5.9	9.9	60
	u3/3	0	0	100
	u3/4	-7.2	-10.0	72

Table 3. Measured and calculated forces in selected cables obtained in static tests.

Ballast setting	Measure point	Measur. force incr. [kN]	Theoret. force incr. [kN]	Accuracy [%]
U34 full load	P4-W7L	644	636	101
	P4-W5L	615	636	97
	P4-W5P	495	575	86
	P4-W7P	326	318	103
U45 full load	P4-W7L	-35	57	61
	P4-W5L	201	238	84
	P4-W5P	315	323	98
	P4-W7P	490	525	93
U3 full load	P4-W7L	162	167	97
	P4-W5L	150	177	85
	P4-W5P	130	163	80
	P4-W7P	77	93	83

4 Dynamic tests

Dynamic tests were carried out based on structural excitation using 5-axle vehicle passes, each vehicle of a total 442 kN weight. The basic tests included passes of a single vehicle with speeds: 10, 30, 50, 70 and 90 km/h and two vehicles with speeds: 10 and 30 km/h. Additional tests included passes of a single vehicle by a threshold with 10 and 30 km/h speed variants. In addition, the frequency and form of natural vibrations were identified using a LMS measurement set and a modal hammer.

In the course of dynamic tests structural response was recorded with the following instruments:

- vertical span displacements by inductive sensors (NOVOTECHNIK): 4 measurement points,
- acceleration by accelerometers (APEK, AV32M37R4 .6U16 and PCB Piezotronics, 393A03): 10 measurement points.

The localisation of all measurement points due to dynamic load tests is presented in Figure 6B.

Figures 8 and 9 display example results, i.e. vertical displacement time history at the u3/34b measurement point during 70km/h vehicle passage and vertical acceleration time history at a1z/45b during 30 km/h vehicle passage with a threshold. The structure detects high resistance to dynamic excitations. The analysis of all vehicle passages allowed to estimate the dynamic increase factor $\varphi = 1.01$.

The results recorded during vehicle passages extended by modal hammer excitations allowed to identify natural vibration frequencies and modes for the bridge. Table 4 presents the obtained natural frequencies of the bridge.

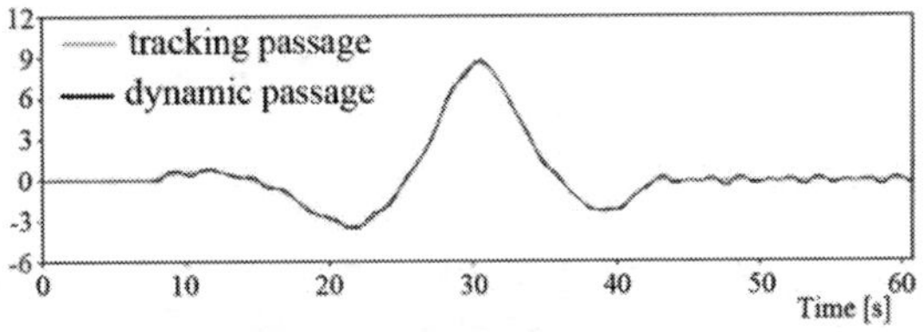

Fig. 8. Vertical displacement time history at point u3/34b excited by a passing vehicle with a 70 km/h speed.

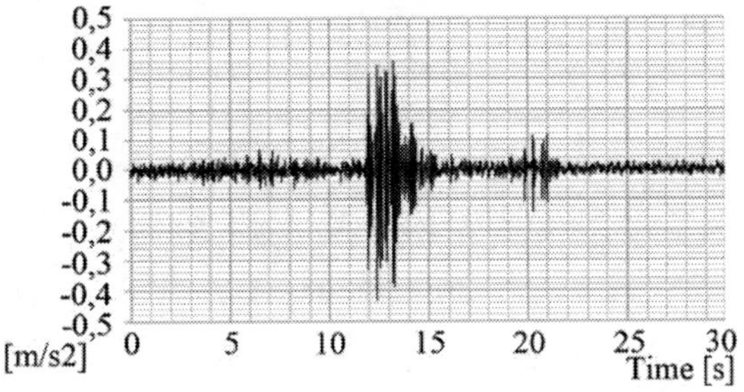

Fig. 9. Vertical acceleration time history at point a1z/45b during a 30 km/h vehicle passage with a threshold.

Table 4. Identified and calculated natural frequencies of the bridge.

Mode shape	Measured frequency [Hz]	Theoretical frequency [Hz]	Accuracy [%]
1	0.31	0.35	89
2	0.49	0.55	89
3	0.74	0.95	78
4	0.91	1.10	83
5	1.21	1.25	97
7	1.50	1.55	97
8	1.67	1.65	101

4 Structure scanning

Nowadays laser scanning is a useful tool for structural geometry assessment during their life cycle. If the scan is made just after competition of e.g. a bridge, the information about its original geometry is stored, to be freely applied in the future [27-33].

At the end of load test performance an entire structural laser scan was completed by Leica P30 (Figure 10). A thorough, spatial image of bridge geometry was achieved by the use of 96 stations and a grid of 1 mm x 1 mm. The position alignment was completed by the least square method with an error not exceeding 5 mm. The created point model of the bridge (Figure 11) can be used to analyse geometry, creating cross-sections (Figure 12) and comparing the scans from two different measurement periods in any point of the structure.

Fig. 10. Laser scanner Leica P30 station.

Fig. 11. Point model of MS-3/DK-16 bridge.

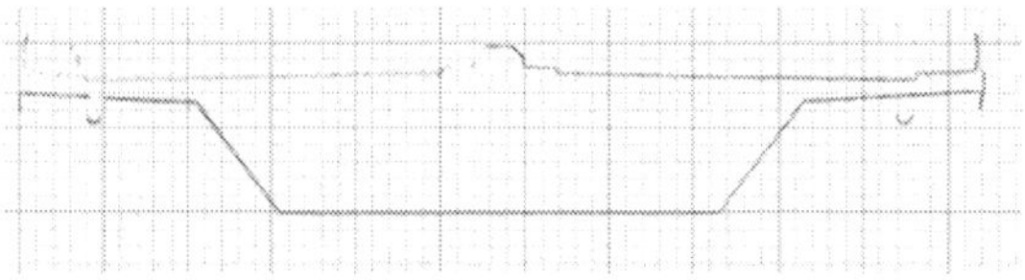

Fig. 12. Example section of bridge deck obtained in laser scanning.

5 Conclusions

The acceptance load tests were carried out in accordance with the program covering both static and dynamic parts. The tests were closed with a positive result: the structural response to ballast and dynamic excitations met the expectations. The spatial structural stiffness is higher than previously assumed in numerical analysis. The maximum registered vertical displacement of the deck during full load of U34 setting is 157.50 mm ± 2.5 mm, i.e. 85% of the theoretical value. The average ratio of elastic values measured to the theoretical ones ranges 70%. Permanent displacements of the spans are up to 2.50 mm ± 0.1 mm, not exceeding the limit values. The structure presents high resistance to dynamic excitations. Laser scan of the structure is a valuable geometry information base to monitor the structural performance in the future.

Acknowledgement

The authors wish to acknowledge the work provided by the others members of the research team: prof. K. Wilde, prof. J. Chróścielewski, prof. M. Rucka, Ł. Filar, R. Kedra, J. Kałuża, K. Makowska, B. Meronk, B. Sobczyk, P. Tysiąc as well as the bridge designer T. Stefanowski for his support.

References

1. D. Collings, A. S. Gonzalez, *Extradosed and cable-stayed bridges, exploring the boundaries*, Proceedings of the Institution of Civil Engineers - Bridge Engineering **166 (4)**, 231–239 (2013) DOI: 10.1680/bren.10. 00058
2. J. Mathivat, W.F. Crozier, FIP Notes **2**, 15-21 (1988)
3. J. Biliszczuk, J. Onysyk, W. Barcik, R. Toczkiewicz, A. Tukendorf, Frontiers in Built Environment **2**, (2017) DOI: 10.3389/fbuil.2016. 00037
4. M. Miskiewicz, Ł. Pyrzowski, K. Wilde, J. Chróścielewski, J. Kałuża, *Load testing of a suspended footbridge in Radom (Poland)*, Shell Structures: Theory and Applications **4** CRC Press/Balkema, 437-440 (2018)
5. M. Miskiewicz, Ł. Pyrzowski, *Load Tests of the Movable Footbridge Over the Port Canal in Ustka*, 2017 Baltic Geodetic Congress (BGC Geomatics), IEEE Xplore Digital Library (2017) DOI: 10.1109/bgc.geomatics.2017.7
6. Ł. Pyrzowski, M. Miśkiewicz, J. Chróścielewski, K. Wilde, *Load Testing of GFRP Composite U-Shape Footbridge*, IOP Conference Series: Materials Science and Engineering **245**, 032050 (2017) DOI: 10.1088/1757-899x/245/3/032050
7. M. Miskiewicz, Ł. Pyrzowski, K. Wilde, *Structural Health Monitoring System for Suspension*

Footbridge, 2017 Baltic Geodetic Congress (BGC Geomatics), IEEE Xplore Digital Library (2017) DOI: 10.1109/bgc.geomatics.2017.8

8. M. Miskiewicz, Ł. Pyrzowski, J. Chroscielewski, K. Wilde, *Structural Health Monitoring of Composite Shell Footbridge for Its Design Validation*, 2016 Baltic Geodetic Congress (BGC Geomatics), IEEE Xplore Digital Library (2016) DOI: 10.1109/bgc.geomatics.2016.48
9. A. Mariak, M. Miśkiewicz, B. Meronk, Ł. Pyrzowski, K. Wilde, *Reference FEM model for SHM system of cable-stayed bridge in Rzeszów*, Advances in Mechanics: Theoretical, Computational and Interdisciplinary Issues, 383–387 (2016) DOI: 10.1201/b20057-82
10. M. Rucka, W. Witkowski, J. Chroscielewski, S. Burzynski, K. Wilde, Bulletin of The Polish Academy of Sciences-Technical Sciences **65 (6)**, 805-813 (2017) DOI: 10.1515/bpasts-2017-0089
11. M. Miśkiewicz, K. Makowska, SGEM2017 Conference Proceedings, ISSN 1314-2704, **17 (22)**, 257-264 (2017) DOI: 10.5593/sgem2017/22/S09 .032.
12. J. Chróścielewski, M. Miśkiewicz, Ł. Pyrzowski, B. Sobczyk, K. Wilde, Composites Part B-Engineering **126**, 153-161 (2017) DOI: 10.1016/j.compositesb. 2017.06.009
13. J. Bien, M. Kuzawa, Archives Of Civil And Mechanical Engineering **15 (4)**, 1046–1060 (2015) DOI: 10.1016/j.acme.2015.05.007
14. T. Siwowski, M. Kulpa, M. Rajchel, P. Poneta, Composite Structures **206**, 814-827 (2018) DOI: 10.1016/j.compstruct.2018.08.048
15. T. Siwowski, D. Kaleta, M. Rajchel, Composite Structures **192,** 555-567 (2018) DOI: 10.1016/j.compstruct.2018.03.042
16. M. Salamak, T. Owerko, P. Lazinski, Architecture Civil Engineering Environment **9 (4)**, 89-97 (2016) DOI: 10.21307/acee-2016-055
17. C. Machelski, M. Hildebrand, Journal of Civil Structural Health Monitoring **5 (1)**, 1-9 (2014) DOI: 10.1007/s13349-014-0074-7
18. T. Mill, A. Ellmann, M. Kiisa, et al., Baltic Journal of Road and Bridge Engineering **10 (1)**, 17-27 (2015) DOI: 10.3846/bjrbe.2015.03
19. P. Olaszek, M. Lagoda, C. Ramon, Structure and Infrastructure Engineering **10 (6)**, 834-842 (2014) DOI: 10.1080/15732479.2013.772212
20. S. Parivallal, T. Narayanan, K. Ravisankar, et al., Strain **41 (1)**, 25-30 (2005) DOI: 10.1111/j.1475-1305.2004.00160.x
21. P. Łaziński, S. Pradelok, Procedia Engineering **161**, 530-537 (2016) DOI: 10.1016/j.proeng.2016.08.684
22. C. Machelski, M. Hildebrand, Journal of Civil Structural Health Monitoring **5 (1)**, 1-9 (2014) DOI: 10.1007/s13349-014-0074-7
23. C. Machelski, Studia Geotechnica et Mechanica **40**, 1-7 (2018) DOI: 10.2478/sgem-2018-0020
24. PN-85/S-10030 Bridge structures. Loads (in Polish)
25. J. Chróścielewski, J. Makowski, W. Pietraszkiewicz: *Statyka i dynamika powłok wielopłatowych. Nieliniowa teoria i metoda elementów skończonych*. IPPT PAN. Warszawa 2004 (in Polish)
26. L. Kwasniewski, H.Y. Li, J. Wekezer, J. Malachowski, Finite Elements in Analysis and Design **42 (11)**, 950-959 (2006) DOI: 10.1016/j.finel.2006. 01.014
27. J. Szulwic, P. Tysiac, A. Wojtowicz, *Coastal Cliffs Monitoring and Prediction of Displacements Using Terrestial Laser Scanning*, 2016 Baltic Geodetic Congress (BGC GEOMATICS), IEEE Xplore Digital Library (2016) DOI: 10.1109/BGC. Geomatics.2016.20
28. K. Bobkowska, A. Inglot, M. Mikusova, P. Tysiąc, Polish Maritime Research Special Issue S1 (93) **24**, 10-15 (2017) DOI: 10.1515/pomr-2017-0015
29. M. Przyborski, P. Tysiąc, E3S Web of Conferences **26**, art. no. 00011 (2018) DOI: 10.1051/ e3sconf/20182600011
30. J. Szulwic, P. Tysiąc, MATEC Web of Conferences **122**, art. no. 04004. (2017) DOI: 10.1051/ matecconf/201712204004
31. K. Bobkowska, J. Szulwic, P. Tysiąc, MATEC Web of Conferences **122**, art. no. 04001. (2017) DOI: 10.1051/matecconf/201712204001
32. P. Tysiac, *Laser Scanning of a Soil-Shell Bridge Structure*, 2018 Baltic Geodetic Congress (BGC Geomatics), 250-253 (2018) DOI: 10.1109/BGC-Geomatics.2018.00054
33. A. Janowski, W. Kaminski, K. Makowska, J. Szulwic, K. Wilde, *The method of measuring the membrane cover geometry using laser scanning and synchronous photogrammetry*, Book Series: International Multidisciplinary Scientific GeoConference-SGEM, ISSN 1314-2704 (2015)

E3S Web of Conferences **63**, 00007 (2018) https://doi.org/10.1051/e3sconf/20186300007
2018 BGC

Assessment of the motorway's impact on agricultural land on the example of the section of motorway A4 Jasień – Bobrowniki Małe

Stanisław Bacior[1,*], *Małgorzata Dudzińska*[2] and *Barbara Prus* [3]

[1]Department of Geodesy, Cadastre and Photogrammetry, Faculty of Environmental Engineering and Land Surveying, Agricultural University in Kraków, Poland
[2]Institute of Geoinformation and Cartography, Faculty of Geodesy, Geospatial and Civil Engineering, University of Warmia and Mazury in Olsztyn, Poland
[3]Department of Spatial Development and Landscape Architecture, Faculty of Environmental Engineering and Land Surveying, Agricultural University in Kraków, Poland

Abstract. The innovative method of assessment of the motorway impact on agricultural land presented in this study allows to estimate all losses related to the directions of this impact. The basis for the determination of losses is the analysis of land use variability and bonitation classes as well as the location of access roads to the land along the axis of the planned motorway. The applied measure of the motorway's impact on agricultural land is the change in land value, at which determination only the diversity of its suitability for agricultural production is taken into account. The developed method of determining the impact of the motorway on agricultural land was presented on the example of the A4 motorway section between the villages Jasień and Bobrowniki Małe. The existing motorway section was evaluated and then the obtained data were compared with the results obtained for the alternative version of the motorway running on the same section.

1 Introduction

The simplified method of assessing the motorway's impact on agricultural land was used to conduct the research in this article. This method is particularly useful during initial motorway route design [2]. This method takes into account only the general assumptions concerning the motorway parameters: the axis of the motorway, the width of the road and the occurrence of protective green belts. It involves the analysis of the course of the motorway axis designed on the registry map. To carry out the analysis properly, additional information is needed on the variability of soil quality on the motorway route, the location of roads and motorway overpasses, areas of agricultural land for which access is associated with crossing the motorway lane and parameters of the lay-out of plots cut by the motorway. The received data are the basis for determining the variability - caused by the construction of the motorway - of these land features, which have a particular impact on their production suitability, and then they are used for a comprehensive assessment of the impact of the motorway construction on agricultural land [16].

The simplified method of assessing the motorway impact on agricultural land enables a comprehensive determination of the impact of the motorway construction on agricultural land including:

- the loss of land taken over for the construction of a road lane,
- decrease in the production capacity of land located near the motorway,
- deterioration in the number of farms cut by the motorway axis.

The value of these plots is the measure of the multidirectional impact of the motorway, at which determination only their production usefulness was taken into account. This value is therefore a measure of valorisation of the agricultural usefulness of land for agricultural production. [1]

For the purposes of the presented study, the existing motorway section has been assessed and next the calculations were repeated for the alternative route of the motorway on the same section, between the villages Jasień and Bobrowniki Małe. Thanks to such research, it is possible to show what effect a change in the motorway route has on the losses associated with the construction of the motorway.

The used method is simplified, what reduces its labour intensity, what allows it to be used while assessing various variants of the route of the motorway still at the stage of the initial investment design. The calculation procedure related to the motorway impact determination has been automated using the developed computer program, which also facilitates its application [3].

* Corresponding author: geomm@poczta.onet.pl

https://doi.org/10.1051/e3sconf/20186300007

2 Characteristics of examined section of the A4 motorway

The simplified method of assessing the motorway impact on agricultural land was applied on the A4 motorway section between the villages Jasień and Bobrowniki Małe. This section is located in the Małopolskie voivodship and in the districts of Brześć and Tarnów. The motorway route runs through the villages Jasień, Mokrzyska, Szczepanów, Sterkowiec, Wokowice, Bielcza, located in the brzeski district and Biadoliny Radłowskie, Łętowice, Bogumiłowice, Wierzchosławice, Gosławice, Rudka, Komorów, Bobrowniki Małe, Bobrowniki Wielkie, located in the tarnowski district. It is 29.317 km long, while in the alternative version designed for the needs of the study, the length is 29.784 km. The motorway designed in the right way should run along the borders of the villages and as far as possible from the larger clusters of buildings. Thanks to this run of the motorway route it is possible to limit to a large extent the growth of costs of the agricultural transport on roads caused by cutting off the land from habitats and to reduce the negative effects caused by car traffic for local residents. The examined section in both versions the existing and the alternative one definitely meets the above requirements. The investment is designed in such a way that it bypass-es the main settlement centres of the village and the larger dense areas of compact buildings at quite a distance, whereas when it is not possible, it crosses them in places where the density of buildings is small.

The examined section of the A4 motorway crosses 85 roads, 25 of which are equipped with motorways viaducts, however, these are in most cases roads that secure transport only to the plots where they are located. In the case of the designed alternative version, the motorway crosses 83 roads, 25 of which would be equipped with via-ducts. In the case of an existing road, motorway viaducts are distant from each other by an average of 1173 meters, while in the alternative version the distance is 1191 meters.

The distance between motorway viaducts is the initial parameter used to assess the motorway's impact on the growth of agricultural transport. A large number of viaducts and short distances between them causes that the impact of the motorway construction on the increase of the distance to the land is relatively low.

When estimating the impact of the test section of the motorway on agricultural land, it was assumed that the road width is close to the maximum and is 70 m in the absence of green protective belts. It was also assumed that the green protective belts have a width of 30 m. Therefore, the total width of the motorway lane will vary from 70 to 130 m depending on the presence of protective belts on one or both sides of it. The adoption of such assumptions makes it possible to estimate the relatively high purchase costs of a wide motorway lane. In practice, however, it happens that they can be reduced according to the actual width of the highway.

In the case of the section under consideration, both in the existing and the alternative version, the occurrence of green protective belts is very different. They occur in all possible configurations. In the case of two-sided belts, which are the least, the width of the motorway is 130 meters, in places where the protective belt is on one side, the road lane width is 100 meters, while for a large part of the section, green protective belts have not been provided and in these places the width of the motorway is 70 meters.

3 Methodology

3.1 Determination of the initial parameters

In order to correctly estimate the output parameters, it is necessary to enter the axis of the motorway on the registry map, and then determine the characteristic points. These are points that are located in places where there is a change in the way of land use, change of the bonitation class and on the borders of the village.

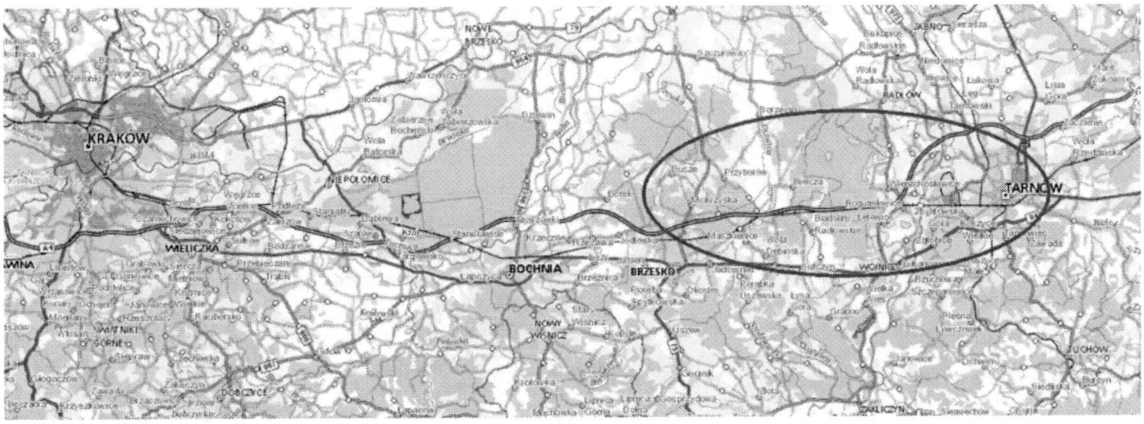

Fig. 1. Location of the analysed section of the A4 motorway (http://mapy.geoportal.gov.pl/imap/?gpmap=gp0)

The next necessary step is to measure the distance between successive points and enter them into the table containing all the initial parameters.

Then, the width of the area with the access via the motorway is determined. This width is determined for all roads cut by the motorway, except the roads that will be equipped with motorway viaducts. To determine this value, it is necessary to know the rules of agricultural transport. This value is directly dependent on the layout of building zones, the layout of roads used for agricultural transport and the borders of the village.

Another important step is to determine the number of agricultural land plots cut by the motorway axis and the width and length of the representative plot. These data are used to calculate the impact of the motorway on the size and shape of the newly created plot.

The last element is to determine whether there are green protective belts along the motorway. On the basis of this information, the width of the motorway and the decrease of the value of land directly adjacent to the road lane are determined. It is assumed that the width of the protective green belt on one side of the motorway is equal 30 meters.

3.2 Area and quality of land taken over for the motorway construction and lying in the range of its impact

The area of land taken over for the motorway construction can be determined as the product of the motorway axis length and its width. It follows that the area depends on the presence of green protective belts and their width. Depending on the presence of the protective belts, the motorway may have a width of 70m, when these belts are not present, and 100m or 130m, assuming that the designed green belt has a width of 30m.

On the basis of the measured sections between the characteristic points it is possible to calculate the areas of different agricultural use and bonitation classes as the product of the sum of the relevant sections and the width of the motorway lane. Thanks to the calculation of the area of bonitation classes, information on soil quality is obtained. Measurement along the motorway axis enables precise determination of the structure of the land use in the area covering 10 to 20% of the motorway lane area. Despite this, such measurement is fully sufficient because the accuracy of estimation of individual surface elements, taken over for the motorway construction depends on the length of the analysed section and for an approximate length of 2 km, we obtain an accuracy of approx. 5 – 10% [1].

3.3 Increase in the access length to the land

The construction of the motorway greatly affects local transport. It cuts many roads and causes the land to be cut off from the settlements. This increases the length of access road to land and, consequently, the increase in transport costs. Very often it happens that the former access is not possible, and the new one is carried out using a roundabout route. Most often, new access roads run along the motorway lane to the nearest motorway viaduct. In this case, the extension of the access road to the land is about half of the distance between roads with viaducts [6, 13].

The determination of the area to which the access will be changed as a result of the construction of the motorway and the increase of the distance to this area, using traditional formulas is an extremely labour-intensive activity. This is mainly due to the necessity to determine the area of land with access via the motorway and the increase of access to these lands. Thanks to the data contained in the initial parameters, the estimation of these parameters is very simple. We obtain them on the basis of the mileage of roads and viaducts and the width of the area with access via the motorway. The values thus obtained are very accurate and best reflect the actual situation [15].

3.4 Deterioration of the plots layout

Another extremely important effect of the motorway construction is the deterioration of the plots layout. The motorway crosses the plots causing a reduction in their area and very often giving them an unfavourable shape, usually plots are divided into two parts. In order to be able to estimate the change of the layout of plots, it is necessary to determine the number of plots cut by the motorway axis and their dimensions. Due to the fact that the most often the plots are divided into two parts, it is assumed that the length of newly created plots and their areas are reduced by half. As a result of the intersection of plots with a motorway, the most often two new plots with a smaller area are created. It follows that the level of changes in plots layouts depends on the ratio of the length of the plot to the width of the motorway [9].

3.5 Determination of the decrease in land value as a result of the motorway construction

The presented method of estimating the values of agricultural lands and their changes caused by the construction of the motorway takes into account only the usefulness of land for agricultural production [3]

This method covers all the basic directions of the motorway's impact on agricultural land, namely:

- taking over the land for the motorway construction,
- an increase in spending on agricultural transport and deterioration of layout of the plots,
- deterioration of the quality of the land in the immediate vicinity of the motorway.

The value of agricultural land calculated using this method is not applicable when determining its market value or determining the value when purchasing the land for construction. This value should be treated as an indicator of the suitability for agricultural production of land, which was determined on the basis of the features of the plot and the farm. The subject of interest is the range of changes in this value due to the construction and unfavourable impact of the motorway.

 https://doi.org/10.1051/e3sconf/20186300007

The decrease in the value of land that followed the construction of the motorway is directly related to the change of those features that have a particular impact on the suitability for agricultural production. It is estimated on the basis of the difference between the value of agricultural land before the motorway construction and after the investment [10, 12].

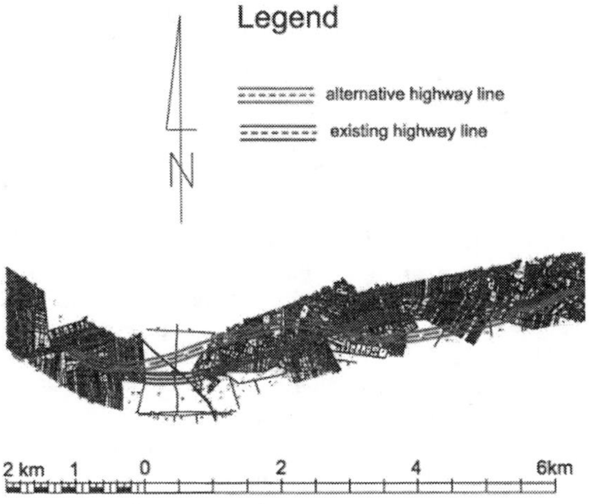

Fig. 2. Location of the existing and alternative section of the motorway (own elaboration)

4 Changes in the characteristics of agricultural land under the impact of the motorway construction

4.1 The area of land taken over for the construction and remaining in the range of the impact of the motorway

The area of land acquired for the construction of the motorway on the section under investigation in the case of the existing version is about 167.17 ha, while in the case of an alternative motorway route this value was estimated at 155.90 ha. This is a fairly large area, taking into account the length of the section. The reason for this is mainly the fact that in both considered versions of the course of the investment there are large fragments, where the green protective belts occur on one side and on both sides of the motorway. Surfaces used for the construction of the motorway and those in the zone of its negative impact include various types of the land use. The largest part of these areas are definitely agricultural land. Their percentage in the area for the existing motorway lane running through the examined villages is 65.9%. While, in the case of the alternative concept, this value is lower and amounts to 54.3%. Often, however, there are cases when the percentage of the agricultural land in the motorway belt is negligible and most often it is caused by the high forest cover of the area through which the motorway runs. Agricultural land area taken over for the construction per 1 km of motorway running through agricultural land for the existing section is equal 7.95 ha, on the other hand, the areas of agricultural land located in the area of its adverse impact are 15.94 ha. For the alternative version of the motorway route, these values are respectively 7.97 ha and 15.97 ha. In both considered cases, the area of land located in the zone of toxic impact of the motorway is twice as large as the area taken over for its construction. The width of the belt, in which there is a harmful effect of the motorway on agricultural land and its surface depends directly on the presence of green protective belts. In the absence of the protective green belts, the negative impact of the motorway can be observed at a distance of up to 90 m from its border and causes a reduction in the quality of agricultural land by an average of 40% [6]. In this case, the width of the toxic impact zone of the motorway is about 180 m, and the assumed width of the motorway itself is about 70 m. The presence of green protective belts on both sides of the motorway causes its width to increase to 130 m, but then the total width of the motorway impact zone is much smaller and is equal approximately 100 m. The presented parameters referring to the area taken over for construction and under toxic impact of the motorway on arable land refer to the assumptions related to the transversal profile of the motorway and regarding the extent of its impact on agricultural land. These values were set in such a way as to highlight the impact of the motorway as much as possible, therefore, the obtained results of this impact may be slightly inflated. An example may be the assumption that the width of the motorway will be equal 70 m, although the minimum width of the motorway with three lanes can be 50 m. However, with the help of the computer software, it is possible very quickly and simply to change the main parameters that define such motorway parameters as: width of the road or the width of green protective belts.

4.2 Cutting off land from the settlements

The axis of the motorway in both cases intersects a large number of roads, which means that a large part of the land will require circular access by motorway viaducts.

The area of land to which the access lengths will increase after the construction of the motorway is, in the case of the existing motorway, 131.72 ha, whereas in the alternative version - 100.68 ha. areas of the land cut off from the habitats by the motorway axis calculated on one kilometre of its course through agricultural land were also estimated. Such conversion eliminates the impact on these areas of the length of a given motorway section and the intensity of occurrence of agricultural land on its route. The motorway section running through agricultural lands on the length of 1 km cuts off 6.26 ha of land from settlements, to which the access lengths will increase, while in the case of the alternative section this area will be equal 5.15 ha.

On the basis of the conducted research, it is possible to state that the increase in the distance to the land caused by the construction of the motorway is directly dependent on the distance between neighbouring viaducts and the number of roads crossed by the section under examination. The increase in distance is bigger when the distance be-tween the viaducts increases and with the great number of roads crossed by the motorway is equal half of that distance.

The increases in distance to land determined by the distances between viaducts are the reason for the increase in expenses for agricultural transport at the moment when they will be accompanied by a large area of land requiring access through roads that will not have viaducts.

4.3 Changes to the layout of the plots divided by the motorway

The existing motorway section on the fragment covered by the study cuts 705 agricultural plots, respectively, while in the alternative concept it is 599 plots. In each case, the motorway most often intersects plots across their lengths, dividing them into two parts.

The average area of plots crossed by the section of the motorway under investigation is about 0.81 ha - in the case of the existing section and 1.06 ha for the alternative section.

The total area of plots for which the land has deteriorated after being crossed by the motorway amounts to 564.37 ha and 640.20 ha in the second concept. This area usually depends on the length of the plots and the length of the motorway section considered, and the size of the percentage of agricultural land in the land use structure in the area covered by the study.

A measure of the deterioration of the plots is the area of plots cut by the motorway calculated for one kilometre of its length running through the agricultural land. In the case under consideration, the area of plots with deteriorated layout referring to one kilometre of the motorway running through agricultural land is 26.83 ha for the existing motorway and 32.73 ha for the alternative version.

In the value of agricultural land caused by the construction of the motorway section, divided into four basic directions of its impact. The obtained values reflect changes in the unit value of the land and the corresponding land areas under the specified direction as a result of the motorway's impact. Table 1 presents the results for the existing section, while Table 2 shows the results for the alternative version in which the course of the investment was changed. For the existing section, the value of land taken over for the motorway construction is 807.7 cereal units / ha per kilometre of the highway running through the agricultural land, while for the section in the alternative concept it is 865.7 cereal units / ha. The land is bought by the investor. The purchase of land occupied by the road lane will cover only about 50% of agricultural land losses caused by the construction of the motorway. Calculated losses in agricultural lands concerning the reduction of their production value may be related to farms with incomplete utilization of their production means, such as farm buildings or agricultural machines. This may be the reason for additional losses incurred by agricultural farms that will be related to the construction of the motorway. Reducing the farm area may have an impact on the number of animals being farmed. Therefore, taking into account the value of land and fixed assets of agricultural production, the actual losses of farms can be estimated at more than twice as much as the decline in the value of agricultural land [2]. On the basis of many researches, it can be concluded that the purchase of land for the construction of the motorway covers only a part of the losses incurred by agricultural farms in connection with the construction of the motorway. However, in reality, the purchase price of land for the motorway construction is even 4 times higher than the average price of agricultural land [13]. In the examined cases, the acquisition of land for the motorway construction amounts to 47.0% of the total loss of land value caused by its construction for the motorway in the existing version, and 52.0% for the alternative version. Such, relatively high result is due to the fact that in both cases there are sections of the motorway for which the green protective stripes have been designed on both sides of the road. Therefore, the area taken over for the motorway construction is quite high and constitutes the largest percentage in relation to the total loss of value of agricultural land related to this construction. Reduction of the quality of lands located near the motorway is equal 558.4 cereal units / ha per kilometre of the motorway, which includes 32.5% of the total reduction in the value of agricultural land, and 537.5 cereal units / ha per kilometre of the motorway (32.3%) in the alternative version of the motorway route. The area and percentage in the total loss of the value of land cut by the motorway axis is directly dependent on the presence of green protection belts. The lack of safety belts means that less land is taken over for the construction of the motorway, and consequently the losses associated with this direction of the motorway are smaller. However, the depreciation of land subjected to the harmful impact of the motorway increases. In the case of the existing motorway section, the total impact of land acquisition for the construction of the motorway and its harmful impact on the land located in its vicinity is approximately 79% of the total.

Tab 1. Decrease in the value of land caused by the construction of the motorway for the existing section (own study).

No	The cause of lowering the value of the land	The value of one hectare of land [cereal unit/ha]		The area covered by the change [ha]	Reduction of income value				
		Before the change	After the change		In the village [cereal unit]	Structure [%]	Per kilometre of motorway [cereal unit/ha]	Per kilometre of motorway section going through agricultural land	Structure [%]
1	The takeover of land for the construction highway	101.65	-	167.17	16993.4	100.0	579.6	807.7	47.0
2	Deterioration in the quality of land located near the motorway	101.65	66.61	335.25	11748.6	69.1	400.7	558.4	32.5
3	The increase in distance of land from the farm due to changes in the communication system	101.65	90.80	131.72	1429.5	8.4	48.8	67.9	4.0
4	Deterioration of layout of plots cut by the highway lane	101.65	91.06	564.37	5976.0	35.2	203.8	284.1	16.5
Total		-	-	1198.51	36147.5	212.7	1233.0	1718.2	100.0

Tab 2. Decrease in the value of land caused by the construction of the motorway for the new section (own study).

No	The cause of lowering the value of the land	The value of one hectare of land [cereal unit/ha]		The area covered by the change [ha]	Reduction of income value				
		Before the change	After the change		In the village [cereal unit]	Structure [%]	Per kilometre of motorway [cereal unit/ha]	Per kilometre of motorway section going through agricultural land	Structure [%]
1	The takeover of land for the construction highway	108.61	-	155.90	16931.5	100.0	568.5	865.7	52.0
2	Deterioration in the quality of land located near the motorway	108.61	74.74	310.35	10512.0	62.1	352.9	537.5	32.3
3	The increase in distance of land from the farm due to changes in the communication system	108.61	96.73	100.68	1195.5	7.1	40.1	61.1	3.7
4	Deterioration of layout of plots cut by the highway lane	108.61	102.43	640.20	3952.6	23.3	132.7	202.1	12.1
Total		-	-	1207.13	32591.6	192.5	1094.3	1666.4	100.0

5 Summary

The simplified method of motorway impact assessment presented in this study takes into account all the most important directions of the motorway's impact on agricultural land and presents this impact in measurable and com-parable units. A very big advantage of the used method is its very low labour intensity, which is caused by the fact that many simplifications were introduced in the assessment of the motorway's impact, thanks to which the scope of obtaining the initial data necessary for the analysis of the course of the motorway axis was largely limited. The automation of calculations with the use of computer software has a huge impact on the speed of analysis [3]. The simplified method of motorway impact assessment is particularly applicable during the initial estimation of the impact of the motorway construction on agricultural land still at the stage of making decisions on the course of the motorway and when assessing the considered variants of the course of the planned sections of the motorway [4]. The research conducted for the needs of this study, to which two sections of the motorway were used, where one is an existing fragment of the motorway, while the other is its alternative version differing in mileage, aims to ilustrate in a special way the application of the simplified method of the impact of the motorway on arable land during the examination various possible variants of the course of this investment. The reduction in the value of land caused by the construction of the motorway on the existing section is equal 1718 cereal units/ha per kilometre of the motorway section, while in the alternative version, which course has been changed, it is equal 1666 cereal units/ha per kilometre of the motorway section. The difference between the de-crease in value in both concepts, in particular, is due to the decrease in value related to the increase in the distance of land from settlements as a result of changes in the transport system and the deterioration of the layout of the plots divided by the motorway lane, which were much more favourable for the alternative motorway concept.

References

1. S. Bacior, S. Harasimowicz, *The impact of the protective green belts on the influence of a motorway on agricultural land,* Surveying Review, **1**, 14-16 (2000)
2. S. Bacior, S. Harasimowicz, *The method of assessing the impact of motorway on the value of agricultural land. The accuracy of estimation of the area of arable land and classes of land,* International Conference of the Polish Society of Agricultural Engineering, 107-120 (2012)
3. S. Bacior, *The impact of motorway section under construction between the towns of Borek Mały and Boreczek on arable land,* Infrastructure and ecology of rural areas, **3**, 5-15 (2010)
4. S. Bacior, *The impact of the motorway on agricultural land on the example of its section Brzozówka – Nowa Jastrząbka,* Acta Scientiarum Polonorum Administratio, **11/2**, 5-14 (2012)
5. S. Bacior, I. Piech, *The impact of line investments on agricultural land on the example of the A4 motorway Żyraków Paszczyna,* Acta Scientiarum Polonorum Geodesia et Desccriptio Terrarum, **12 (1)**, 19-28 (2013)
6. J. Curzydło, *Drogi i autostrady, a skażenie motoryzacyjne,* Aura, **5**, 7-8 (1994) (in Polish)
7. R. Cymerman, *Arangement agricultural method of valuation of agricultural areas,* Valuation, **1**, 12-13 (1992)
8. P. L. Gilbert, A. Steinherr, *Private Finance for Public Infrastructures: in search of a new Framework,* EIB, **23**, 86-87 (1994)
9. S. Harasimowicz, *The main directions of the motorway impact on the income possibilities of farms,* Science notebooks of Agricultural University in Krakow, 57-62 (1997)
10. S. Harasimowicz, *Evaluation of the motorway impact on the agricultural land,* Surveying Review, **6**, 6-12 (1998)
11. S. Harasimowicz, *Evaluation and organization of the territory of a farm,* Script of the Agricultural University in Krakow (2002)
12. E. Quinet, *Issues in Building the European High-Speed Train Network,* Journal of Infrastructure Systems, **3**, 54 (1997)
13. W. Wilkowski, *Multi-criteria method for assessing the impact of motorway on farms,* Scientific work of the Warsaw University of Technology, **33**, 5-19 (1995)
14. M. Żak, *The issue of valuation of real property located in the belt of construction the motorway network,* Science notebooks of Agricultural University in Krakow, **396**, 17-22 (2002)
15. S. Bacior, *Auswirkungen eines Autobahnbaues auf Ackerböden - am Beispiel des Abschnitts der Autobahn A-4 zwischen den Städten Brzesko und Tarnów in Polen,* Berichte Uber Landwirtschaft, **95/2**, 1-20 (2017) (in German)
16. S. Bacior, B. Prus, *Infrastructure development and its influence on agricultural land and regional sustainable development,* Ecological Informatics, **44**, 82-93 (2000)

Analysis of the impact of construction of selected water reservoirs on the surrounding environment made on the basis of satellite data

Boguslawa Kwoczynska [1,*], *Izabela* Piech[1], and *Rafal* Wozniak[2]

[1] University of Agriculture in Krakow, Balicka 253a, 30-198 Krakow, Poland
[2] Krakow Office of Geodesy and Agricultural Areas, Gazowa 15, 31-060 Krakow, Poland

Abstract. The authors aimed at showing in the publication the impact of the construction of selected water reservoirs in Poland on the surrounding environment, basing on the satellite imagery. For this purpose, for four test objects, i.e. the Dobczyce Reservoir, the Klimkow Reservoir, the Czorsztyn Reservoir and the Domaniowski Reservoir analyzes were carried out concerning the changes in the structure of land use for the state before and after reservoir construction, and indicators such as NDWI, SAVI and TSAVI were calculated. In the case of the analysis of changes in the land use structure, the direction of these changes was determined first of all, and for SAVI and TSAVI indices, the percentage change in their value and the direction of these changes were calculated.

1 Introduction

Water is the most important substance for human life. Without it, it's hard to imagine life on Earth, the world of plants or animals. People often waste a lot of it without thinking about the state of resources of this life-giving substance in our country. According to the National Water Management Board, based on data from many years (1951 - 2006), average stocks of flowing waters in Poland amounted to 61.9 billion of m3 / year, of which the own resources amounted to 54.3 billion of m3 / year [9]. In addition, there are 2856 lakes in Poland with the area larger than 10 ha and 99 retention reservoirs with the capacity of more than 2 million of m3. In total, it amounts to 21.7 billion of m3 [1]. Trying to look at the distribution of water resources in the country, it can be noticed that the most of water is located in the southern Poland.

An important issue in addition to the availability of water, understood as the state of its resources, is its quality. According to the European Environment Agency (EEA), in the case of Poland, the ecological status of 70 - 90% of water resources is assessed as worse than good [2].

Bearing in mind the above data, the authors asked themselves what to do in such a situation, to take better care of Poland's water resources. A certain remedy for this state of affairs may be the retention reservoirs. The artificial water dams existing currently retain only 6% of the average annual outflow, which is not enough to protect the country against floods and droughts. The physico-geographical conditions in Poland allow for the storage of nearly 15% of the annual outflow [1]. There are about 850 water damming facilities in Poland [3].

Water reservoirs in Poland are located mainly in the south of the country, on mountain rivers. In connection with the existing reserve, in the near future construction of further water reservoirs in Poland may be expected. Especially when it comes to the so-called small retention reservoirs. According to [4] and [3], retention is the phenomenon of temporary water holding in the river basin. On the surface of the terrain, the water is accumulated in lakes and water reservoirs, it fills riverbeds valleys and remains in the field hollows, puddles, field furrows.

Taking into account the above-mentioned factors, it is necessary to consider the impact of water reservoirs on the surrounding environment. In the publication, the authors focused primarily on an attempt to analyze this impact basing on data from different sensors, installed on the air platforms (airborne scanning, aerial cameras) and satellite platforms (multispectral scanners).

Applications of satellite imagery in assessing changes occurring in the natural environment for many years is presented both in Polish [5-9] and foreign [10-15] literature. However, only a few were involved in the application of remote sensing or photogrammetry to study the impact of building water reservoirs on the environment [14, 16, 17].

2 Research work

2.1 Satellite remote sensing

The first human space missions, such as the American missions Mercury, Gemini or Appolo, as well as the Soviet and then Russian Soyuz, can be considered the

* Corresponding author: rmkwoczy@cyf-kr.edu.pl

beginning of the satellite remote sensing. Initially, the pictures of our planet were made with the use of handheld cameras that did not provide good quality data [18]. In the case of the Mercury mission - Maurer, Ansco, Robot Recorder or Hasselblad cameras were used for photographing [5].

A milestone in the development of satellite remote sensing was the mission Apollo 9 (March 3 - 13, 1969), in which the experiment consisting in taking multispectral photographs was performed for the first time. For this purpose, 4 Hasselblad cameras synchronized with each other were used, each of them with a film sensitive in a different range of light waves [5]. As a result of this experiment, photographs were taken in panchromatic, green, infrared channels and in the spectroscopic channel. The terrain resolution of the data obtained in this way was 100 - 130 m [5]. Initially, satellite images were used similarly to airborne photographs - for the interpretation of elements of the natural environment based on the assessment of the photo features. A certain novum has become the use of so-called Additive Monitors, thanks to which it was possible to assemble an RGB image based on individual scenes in given spectral channels.

After the first manned missions, two Earth observation missions were carried out in parallel in the United States: it was the Landsat mission started in 1972 (continues to this day), and the first orbiting space station SkyLab (1973 - 1974) [5]. The SkyLab station equipment was particularly interesting. It was equipped with a Multispectral ITEK camera (6 cameras equipped with filters: colored, panchromatic, infrared, false-color), high resolution photo camera, infrared spectrometer, multispectral scanner (13 channels), radiometer working in the K band and the second working in the L band [5].

At the same time, advanced works on the use of satellite images in the economy were continued in the Union of Soviet Socialist Republics (now Russia) (Vostok 2, Voschod and Soyuz, Salut). These works also did not avoid Europe, where the French SPOT mission (Systeme Pour l'Observation de la Terre), which began in February 1986 is worth mentioning.

Also the European Space Agency established in 1975, carried out remote sensing missions. One of the first was the Envisat satellite launched in 2002. Its successor was the Sentinel satellite, which has been launched since 2014. There are currently 5 satellites of this mission in space - 2 equipped with MSI multispectral scanners and 2 with SAR radars. Satellites of the Sentinel 3 mission are equipped with various sensors, such as MWR, OLWI, STSLR, and SRAL.

At present, the market of commercial satellites which provide high resolution data is also developing. The beginning of this type of companies should be seen in 1992, when Russia began selling images from KVR-1000 and KFA-3000 cameras (resolution 2 - 3 m). In 1999, the mission of the Ikonos-2 satellite began. In 2001, DigitalGlobe launched the QuickBird-2 satellite. The satellites of these missions began to provide data at resolutions of about 0.5 m panchromatically and about 1 m multispectrally [18]. An example of a modern commercial satellite is WorldView-4, which has a panchromatic resolution of 0.31 m and a multispectral 1.24 m [19].

2.2 Landsat mission (1972 - present)

For the needs of the research, LANDSAT imaging was used. The LANDSAT mission began in 1972 and continues to this day. Currently (May 2017), there are two satellites of this mission in orbit - Landsat 7 and 8. Table 1 provides basic information on all hitherto LANDSAT missions. The graphics view with the image of particular vehicles are shown in Figure 1.

Fig. 1. Landsat mission satellites. Top row from the left: Landsat 1, 2, 3, 4. Bottom row from the right: Landsat 5, 7 i 8. Source: (USGS 2017) [access: 05.06.2017]

Table 1. Basic information on the Landsat mission. Source: (USGS 2017)

Satellite	Duration	Spectral channels
Landsat 1	23.07.1972 – 6.01.1978	3 RBV scanner and 4 MSS
Landsat 2	22.01.1975 – 27.07.1983	3 RBV scanner and 4 MSS
Landsat 3	5.03.1978 – 7.09.1983	2 RBV scanner and 5 MSS
Landsat 4	16.07.1982 – 14.12.1993	4 MSS scanner and 7 TM
Landsat 5	1.03.1984 – 5.06.2013	4 MSS scanner and 7 TM
Landsat 6	5.10.1993 (did not reach the orbit)	8 ETM scanner
Landsat 7	15.04.1999 – currently	8 ETM scanner +
Landsat 8	11.02.2013 – currently	9 OLI scanner and 2 TIRS

When describing the capabilities of a given remote sensing system, we often use the following four terms: spatial resolution, spectral resolution, temporal resolution, and radiometric resolution. The first of these terms means the size of a single terrain pixel, or the smallest object possible to be visualized by the sensor. The second is the number of spectral channels and the number of magnetic wavelengths for each channel. The third concept concerns the length of the revisit time over the same point on the Earth. Finally, the last, fourth concept applies to the sensitivity of the sensor, i.e. the number of bits that can be registered. For the 8-bit resolution, it is possible to register values from the range 0 – 255 (28 = 256) [15].

Table 2 presents sample data of sensors installed on the Landsat satellite 8.

 https://doi.org/10.1051/e3sconf/20186300008

Table 2. Spectral channels with their resolutions fixed on the Landsat 8 satellite. Source: (USGS 2017)

Channel (sensor)	Wave length [µm]	Spatial resolution [m]
Channel 1 (OLI)	0.43 – 0.45	30
Channel 2 (OLI)	0.45 – 0.51	30
Channel 3 (OLI)	0.53 – 0.59	30
Channel 4 (OLI) - Red	0.64 – 0.67	30
Channel 5 (OLI) – nIR	0.85 -0.88	30
Channel 6 (OLI) – SWIR 1	1.57 – 1.65	30
Channel 7 (OLI) – SWIR 2	2.11 – 2.29	30
Channel 8 (OLI) – PAN	0.50 – 0.68	15
Channel 9 (OLI) – clouds	1.36 – 1.38	30
Channel 10 (TIRS) – TIRS 1	10.60 – 11.19	100
Channel 11 (TIRS) – TIRS 2	11.50 – 12.51	100

2.3 Characteristics of research objects

For the purpose of performing the necessary analyzes, being the basis for modelling the impact of a water reservoir construction on the surrounding environment, six test objects were selected: Dobczycki reservoir, Czorsztynski reservoir, Domaniowski reservoir, Klimkowski reservoir, Sieniawa reservoir and Świnna Poręba reservoir (Fig. 2).

As the guiding principle of the selection of test objects, it was assumed that selected artificial reservoirs should not be built earlier than in 1970. This was suggested by the fact that the mission of American remote sensing satellites Landsat, which provide data useful for conducting analyzes has been underway since the 1970s.

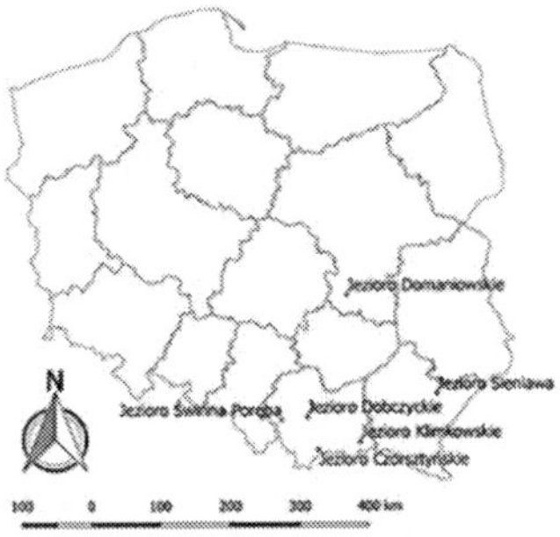

Fig.2. Map of the location of test objects against the background of the country. Source: own elaboration.

Dobczycki Reservoir
The first of the analyzed reservoirs is Dobczycki Reservoir. It is located in the Myslenice District, in the Malopolskie Voivodeship. It was built in 1986 [3] on the Raba River. The total area of the reservoir is equal 1070 ha, and its total capacity amounts to 125.0 million m3. The main task of the Dobczycki Reservoir is to supply the city of Krakow with drinking water, therefore, the retention function is completely excluded. In addition, it has a flood control function and in a limited range energy function.

Czorsztynski Reservoir
The second reservoir is the Czorsztynski Reservoir put into use in 1997. This object is located in the Nowy Sacz district, in the Pieniny area. The area of the reservoir is equal 1234 ha, with the total capacity of 231.9 million m3. The reservoir has primarily energy functions, but it is also a part of the flood protection of the Dunajec Valley. The lake is open to tourists and anglers.

Klimkowski Reservoir
Klimkowski Reservoir (Klimkowka) was created by building a dam on the Ropa River. Located in the Malopolskie Voivodeship, in the Gorlice district. The reservoir was put into service in 1994. It is much smaller than the previous reservoirs - its area is equal 304 ha, and the capacity amounts to 43.5 million m3. Its main tasks are flood protection and municipal water supply. To a lesser extent, it has energy and recreational functions.

Sieniawski Reservoir
It is also known as the Besko water reservoir. The object is located in the Podkarpackie Voivodeship, in the district of Krosno. It was created by building a dam on the Wislok river in 1978. The surface of the created reservoir is equal 126 ha and its total capacity amounts to 13.7 million m3. The main purpose of the reservoir is the flood protection of the Wislok Valley and the supply of water for the Water Treatment Plant in Sieniawa.

Domaniowski Reservoir
The Domaniowski reservoir is located in the Masovian Voivodeship, in the Przysucha district on the Radomka River. It was created in 2001. Its total area is equal 500 ha, and the capacity amounts to 9.9 million m3. The main tasks for this reservoir are flood protection and recreational functions.

Swinna Poreba Reservoir
The last object is the Świnna Poręba Reservoir. It is located on the Skawa river, in the Wadowicki district, in the Małopolskie voivodship. In 2010, its construction was completed and at the same time its filling started. The planned area of this object is equal 1040 ha, and its planned total capacity is equal 161 million m3. The reservoir's task will be flood protection of the Skawa river valley and supply of water for municipal purposes [3].

3 Research methodology

3.1 Unattended classification method

To study the phenomenon of the impact on the environment of newly constructed water reservoirs, the best suited are the multispectral imaging and various methods for classifying these data. The research focused mainly on unattended methods, and more specifically on K-means clustering (k-means, centroid algorithm) and ISODATA.

3.1.1 Method K – means clustering (k – means, centroids algorithm)

For this algorithm, the K value is determined first, denoting the number of clusters (groups of objects with similar properties or values). Then the preliminary initialization of the values of clusters, i.e. the calculation of their average values is performed. In the next steps, each point is assigned to a given cluster by iteration method, and its average value is recalculated. Thanks to this procedure, the average "wanders" to the center of the cluster.

The computational complexity of the presented algorithm can be calculated using the formula (1):

$$O=n\bullet K\bullet l\bullet d \quad (1)$$

where: n – number of points, K – number of clusters, l – number of iterations, d – number of attributes.

In the case of implementation of this algorithm for the purposes of classification of remote sensing data, the number of clusters, or the concept of a cluster is extremely intuitive, because in this case clusters mean a given form of land use. A certain drawback associated with the classification is the necessity to declare the number of clusters before the algorithm is executed.

In K – means method previously, the optimal number of clusters should be assumed. This can be done by trying to determine various forms of land development using different combinations of spectral channels forming on the RGB composition, highlighting the given terrain form. Then, in order to optimize the process of image processing, the large scenes from remote sensing satellites can be cut according to the research area. The last recommendation applies to the maximum number of iterations, however, the selection of the optimal value must be preceded by a previous examination and comparison of effects for different values.

3.1.2 ISODATA

The second method of unsupervised classification used in the carried out research is the ISODATA method (Iterative, Self-Organizing, Data Analysis Technique aYa). Unlike in the K-means method, it is not necessary to provide the number of clusters at the beginning of the analysis. The algorithm combines and divides clusters.

In the first step, the centers of clusters are arranged randomly, and the points are allocated on the basis of the closest distance to the centers. Then, the standard deviation is calculated for each of the clusters. If its value is greater than the assumed distance, the cluster is divided. If the deviation is smaller than assumed, then two clusters are combined into one. After this procedure is finished, another iteration is performed.

The entire algorithm is aborted if one of the three situations occurs: the average distance between cluster centers is smaller than the assumed value; the size of changes in the distance between centers in subsequent iterations is smaller than assumed and finally the maximum number of iterations is reached [20].

The ISODATA algorithm has the advantage over K-means that the number of clusters is automatically selected.

While analyzing changes in the land use structure, which may be affected by the construction of a water reservoir, the following methodology was used:

- Determination of the area to be analyzed - it was assumed that the area would be within 500 m from the shoreline of the reservoir. This is caused by the desire to focus only on the impact of the water body itself, so areas located further could falsify the results of research.
- Cutting the raster files containing individual spectral channels.
- Performing the classification unattended by the k - means method and the ISODATA method (in the case of the first method, additional post-processing was necessary).
- Two analyzes were performed for each research area - for the state before construction and after the construction of the water reservoir.
- Preparation of the statement of the results in tabular form.

3.2 Indices: NDWI, SAVI and TSAVI

Besides the NDWI index, or the NDVI index, which is popular in remote sensing, the subject literature mentions other useful indices. Some of them can be mentioned here, for example PVI (Perpendicular Vegetation Index), RVI (Ratio Vegetation Index), SARVI (Soil Adjusted Ratio Vegetation Index) or SAVI (Soil Adjusted Vegetation Index) and TSAVI (Transformed Soil Adjusted Vegetation Index) [12]. In the course of further research on the impact of the construction of water reservoirs on the environment, the authors focused on using the last two indices to assess this impact.

3.2.1 NDWI

The remote sensing of the natural environment has many applications. One of them is monitoring of the plants condition or estimating the size of crops and determining the amount of biomass [11]. For this purpose, various indices calculated on the basis of the values of the respective spectral channels are used. One of them is NDVI (The Normalized Difference Vegetation Index) [21]. This index uses the contrast between the reflection in the near-infrared band (NIR), and absorption in the red band (RED). It can be expressed using the formula (2):

$$NDVI= (NIR-RED)/(NIR+RED) \quad (2)$$

A certain drawback of this index is the different ability to penetrate vegetation by the both bands. Close infrared can penetrate up to 8 layers of leaves, while the red channel only 1 or less [22]. Bearing in mind these defects of the NDVI index, and wishing to focus more on the detection of liquid water in vegetation, [17] proposed a new index called NDWI (Normalized

 https://doi.org/10.1051/e3sconf/20186300008

Difference Water Index). Its calculation uses two infrared spectral channels with wavelengths close to 0.86 and 1.24 μm. This index is given by the formula (3):

$$NDWI= [B(0.86)-B(1.24)]/[B(0.86)+B(1.24)] \quad (3)$$

Where B(0.86) and B(1.24) mean spectral channels with wavelengths given in brackets. This index takes values from -1 to 1, where the higher the value, the higher the water content. According to [17], open soils in most cases have negative values, but close to 0.

The NDWI index can have many applications, ranging from drought monitoring [14], to analysis of water stress in plants, etc. However, according to [17], NDWI also has some disadvantages that should be kept in mind. In addition to water in plants, the value of this index is also influenced by factors such as soil, pesticides or plant diseases. For this reason, other tools should be considered in addition to using the NDWI index.

The authors adopted the assumption that the construction of a water reservoir will have an impact on the surrounding environment also by changing soil moisture and water availability for plants. Therefore, it was decided to check the change in the value of the index before and after the construction of the reservoir.

Similarly to the analysis of changes in the structure of land use, also here it was decided to limit the area to 500 m from the borders of the reservoir. For the such determined area, an NDWI map was developed and assumption was made to calculate the mean values of the index for selected test fields located in different places around the reservoir.

3.2.2 SAVI

SAVI was calculated to minimize the impact of the brightness of soil images on the results of vegetation indices [23]. Its calculation uses data from two spectral channels: near infrared (NIR) and red (RED). Furthermore, the parameter L was added as a modification of the previous (NDVI) index. It depends on land cover, and for soil not covered with vegetation, the value of 1 is assumed, whereas for full coverage 0 [23]. For average conditions, the value L = 0.5 should be taken [13]. The index takes values from - 1 to 1, whereas, the closer to 1, the better is the vegetation. In order to calculate the SAVI index, we use the formula (4):

$$SAVI= [(1+L)\cdot(NIR-RED)]/(NIR+RED+L) \quad (4)$$

Where NIR and RED are appropriate spectral channels (for Landsat 8 they are respectively channels 5 and 3, and for Landsat 5 channels 4 and 2), and the parameter L, as already mentioned, assumes a value of 0.5.

3.2.3 TSAVI

The TSAVI index was developed as a modification of the formula (4). The main idea of the creator of this index was the desire to closely correlate the vegetation index with such indices as LAI (Leaf Area Index) or APAR (photosynthetically active radiation absorber by the canopy) [24]. This index is given by the formula (5):

$$TSAVI=[a\cdot(NIR-a\cdot RED-b)]/(RED+a\cdot NIR-a\cdot b) \quad (5)$$

where a = 0.96916 and b = 0.084726. TSAVI takes values from 0 for soil not covered by vegetation up to 1 for a very high value of LAI, ie rich vegetation [24].

As in the previous cases, analyzes of the state before and after construction of water reservoirs were performed to determine their impact on the values of indices. Analyzes were made as before for four test objects, i.e. Dobczycki, Czorsztynski, Klimkowski and Domaniowski reservoirs. The same test fields as for the NDWI analyses were used. When calculating the SAVI value, for the L parameter the value L = 0.5 was assumed. All results were collected in the appropriate tables.

4 Results

From the initially selected reservoirs, the Sieniawski Reservoir was excluded (due to the small area) and the Świnna Poręba Reservoir (because of its lack of filling - the lake has not yet reached the planned surface, so it is not possible to talk about its impact here, at least about the study of the impact on the environment).

Initially, according to the presented methodology, it was planned for each reservoir to perform analyzes before filling and for 2016 using K-means and ISODATA methods. However, after visual analysis of the results obtained with the latter method, it was decided to exclude it from the further analyzes.

Also, after analyzing the results for the Czorsztynski Reservoir, it was considered that it should be excluded, because incorrect results could have a significant negative impact on the results of the entire experiment. An example of incorrect classification for the Czorsztynski Reservoir is shown in Figure 3. Finally, three reservoirs were taken into consideration: Dobczycki, Klimkowski and Domaniowski.

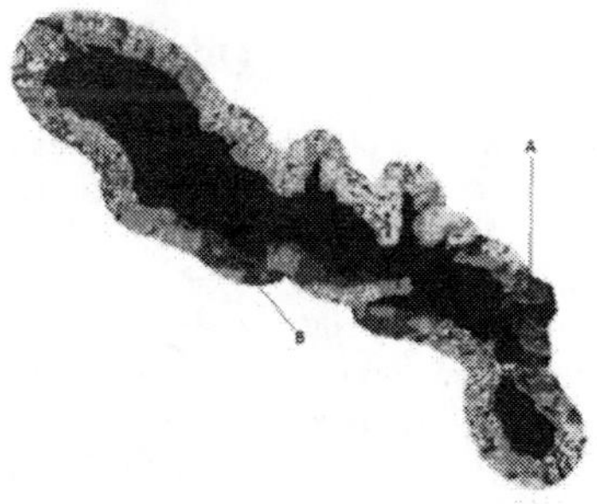

Fig. 3. Erroneous classification of the unattended Czorsztyn Lake. Source: own elaboration.

The method K– means used here, gave incorrect classification. Two areas are noteworthy – A, being a hill with the Czorsztyn castle, erroneously classified as a water surface area, and B – area of the beach at the lake.

https://doi.org/10.1051/e3sconf/20186300008

Table 3. The summary of the results of analyzes of changes in the structure of agricultural land before and after the construction of water reservoirs. Source: own study.

Type of use	Dobczycki Reservoir			Klimkowski Reservoir			Domaniowski Reservoir		
	1984 [ha]	2016 [ha]	Change %	1984 [ha]	2016 [ha]	Change %	1994 [ha]	2016 [ha]	Change %
Forest areas	413.46	622.98	50.67	376.38	415.80	10.47	287.64	278.37	-3.22
Urbanized areas	24.39	175.59	619.93	11.52	20.16	75.00	0.00	3.06	-
Agricultural areas	592.56	272.70	-53.98	97.47	114.93	17.91	248.13	186.75	-24.74
Meadows and pastures	498.87	225.00	-54.90	178.02	87.57	-50.81	104.04	33.75	-67.56
Thicket and wasteland	32.22	260.28	707.82	9.54	34.47	261.32	51.75	124.11	139.83
Water	823.32	-	-	212.85	-	-	277.92	-	-

4.1 Analysis of the land use changes around selected water reservoirs

For each analysis, the necessary step was to identify spectral homogeneous areas identified in the unattended classification process. To accomplish this, two sources were used – orthophotomap (mainly for the current state of use) and a color combination, created by assigning appropriate spectral channels to RGB colors. Using the properties of certain objects, which consists in differentiation of the absorption of different wavelengths of light, it is possible to create images that enhance some forms of space management in a simple way through an appropriate combination (Fig. 4).

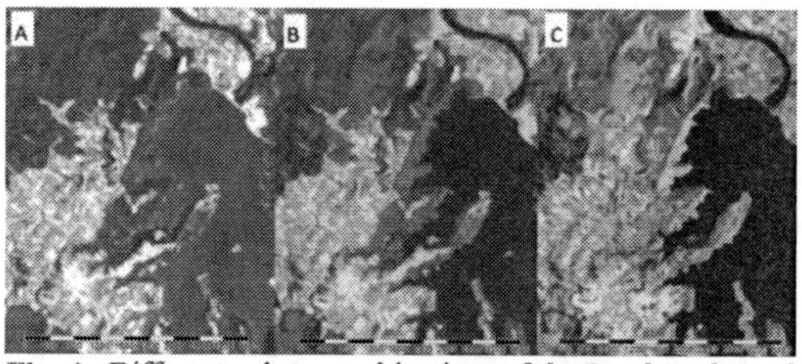

Fig. 4. Different colour combinations of the Landsat channels: A – 321, B – 432, C – 541. Source: own elaboration.

Table 3 presents the results of analyzes of changes in land use for the three analyzed areas.

Analyzing the obtained results, it can be noticed that in the case of both Dobczycki and Klimkowski Reservoirs, there was an increase in the area occupied by forests, respectively by 50 and 10% in relation to the initial state. Only in the case of the Domaniowski Reservoir there was a loss of forest areas, but this is about 3%, and in the opinion of the authors, this can be considered as an error related to the classification.

At the same time, in each case an increase in urbanized areas was noted, the greatest around the Dobczycki reservoir (nearly 620%), but this can be explained by the proximity of the city in relation to Krakow and increased construction pressure on the part of the inhabitants of this city. Another interesting analysis is the reduction in the area of land occupied by meadows and pastures (by about 50 - 67%), and the reduction of agricultural land. Only in the case of the Klimkowski reservoir, this area has slightly increased (nearly 18%). Along with the reduction of land area related to agriculture, the area of wastelands and thicket increases. The largest percentage share was recorded in the case of the Dobczycki Reservoir and it amounted to over 700% in relation to the initial state.

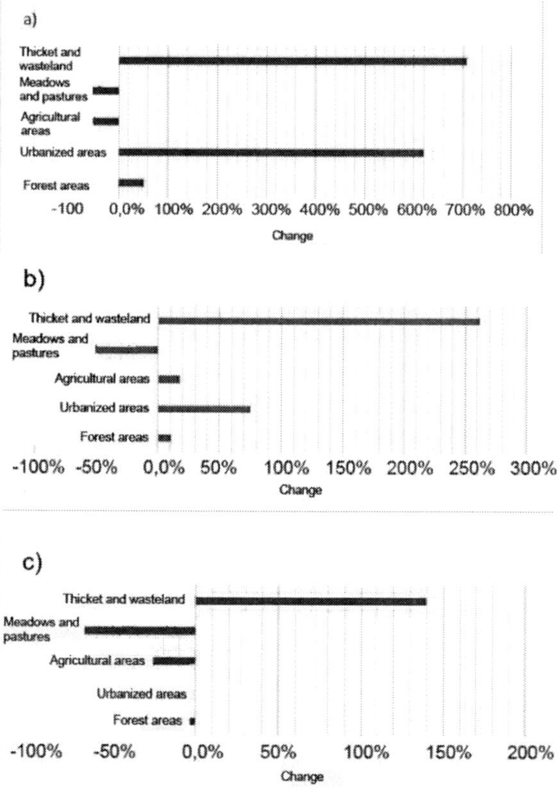

Fig. 5. The dynamics of changes in land use around water reservoirs a) Dobczycki Reservoir; b) Klimkowski Reservoir; c) Domaniowski Reservoir. Source: own elaboration.

https://doi.org/10.1051/e3sconf/20186300008

In order to better present the data, the results were provided in the form of diagrams (Fig. 5), where the dynamics of changes are presented, and the selected land cover maps are show on the drawings 6-8.

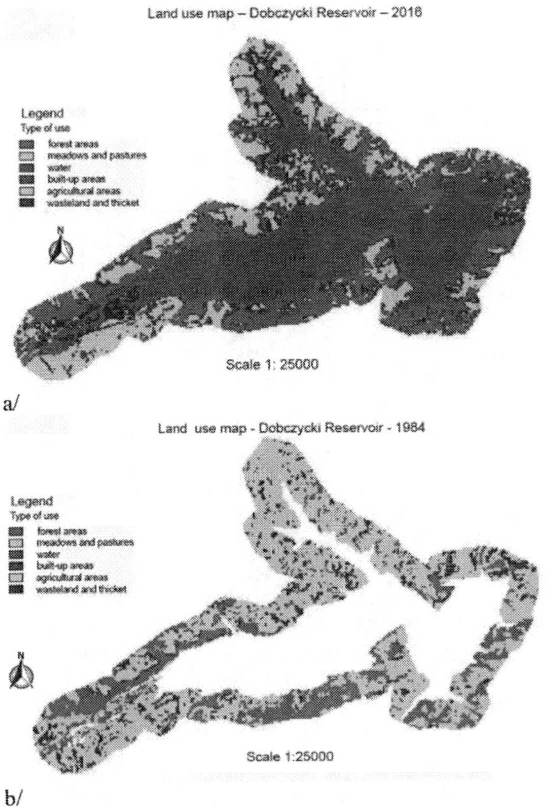

Fig. 6. The map of land use around the Dobczycki Reservoir made on the basis: a/ Landsat 8 satellite from 2016, b/ Landsat 5 satellite from 1984.

4.2 Analysis of the NDWI index changes for selected water reservoirs

In order to examine the changes in the NDWI index, calculations were made for 3 water reservoirs: Dobczycki, Klimkowski and Domaniowski Reservoirs. Two analyzes were performed for each of them: the state before the construction of the reservoir and the state after its construction. Thanks to this, it was possible to determine the change in water content in plants. For each reservoir, test fields located in the test area were selected, i.e. in the buffer zone of 500 m from the reservoir borders (Fig. 9).

Not the same number of test fields resulted from the size of reservoirs and their shapes.

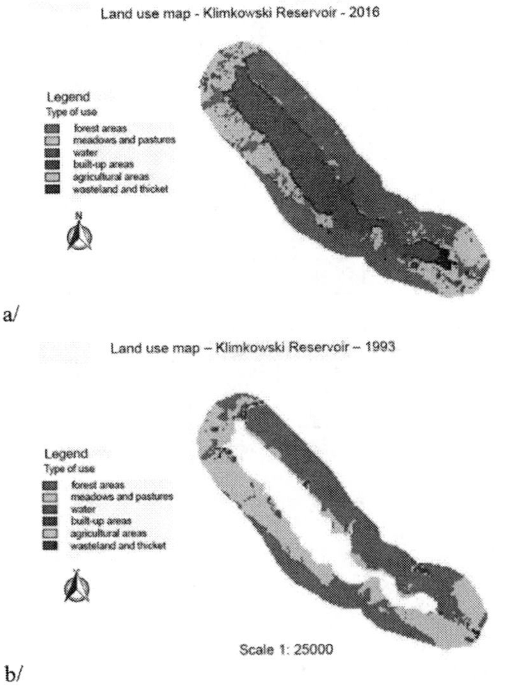

Fig. 7. The map of land use around the Klimkowski Reservoir made on the basis: a/ Landsat 8 satellite from 2016, b/ Landsat 5 satellite from 1993.

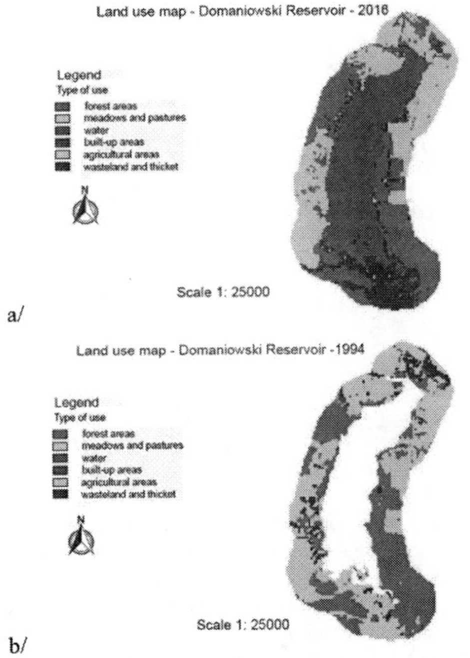

Fig. 8. The map of land use around the Domaniowski Reservoir made on the basis: a/ Landsat 8 satellite from 2016, b/ Landsat 5 satellite from 1994.

 https://doi.org/10.1051/e3sconf/20186300008

Table 4. Changes in the value of the NDWI index for the examined objects. Source: own elaboration.

	Dobczycki Reservoir				Klimkowski Reservoir				Domaniowski Reservoir			
field	Area [ha]	NDWI 1984	NDWI 2016	Change %	Area [ha]	NDWI 1993	NDWI 2016	Change %	Area [ha]	NDWI 1994	NDWI 2016	Change %
1	3.4556	-0.4173	-0.2972	28.76	7.7730	-0.3883	-0.2953	23.95	13.0585	-0.3337	-0.3015	9.64
2	1.8313	-0.3921	-0.2819	28.12	7.3659	-0.4291	-0.2842	33.78	5.4163	-0.2631	-0.2336	11.20
3	3.7416	-0.2676	-0.2755	-2.96	5.6728	-0.5181	-0.3979	23.21	9.0832	-0.2963	-0.2010	32.16
4	9.1433	-0.3842	-0.2809	26.88	7.3664	-0.4506	-0.2569	42.99	12.1085	-0.2993	-0.1991	33.48
5	13.3009	-0.3771	-0.3325	11.83	10.0902	-0.4865	-0.3603	25.95	9.8110	-0.2447	-0.1893	22.64
6	4.0952	-0.4269	-0.2972	30.37	-	-	-	-	16.2587	-0.4027	-0.2532	37.12
7	5.2073	-0.4574	-0.3510	23.25	-	-	-	-	-	-	-	-
average		**-0.3889**	**-0.3023**	**20.89**		**-0.4545**	**-0.3189**	**29.97**		**-0.3066**	**-0.2296**	**24.38**

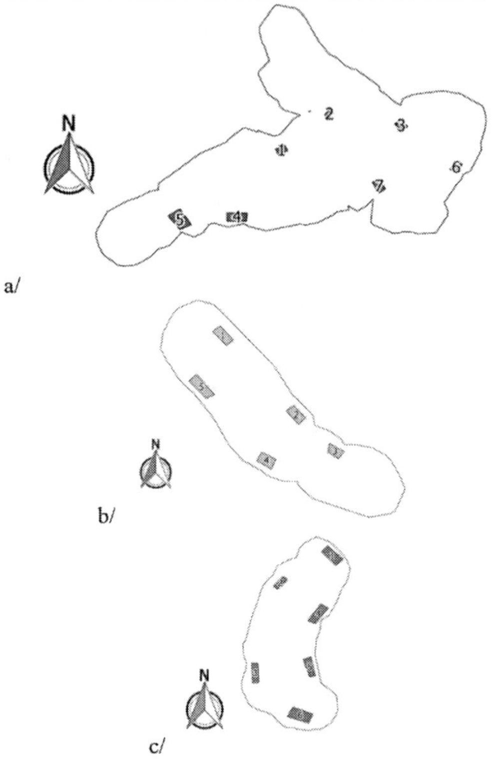

Fig. 9. The map of the test fields locations for the object: a/ Dobczycki Reservoir, b/ Klimkowski Reservoir, c/ Domaniowski Reservoir. Source: own elaboration.

For each of the test fields, the average value of the NDWI index for the state before and after the reservoir construction was calculated and the area of each field was calculated. The results are summarized in Table 4. Moreover, the list includes the change in the value of the index as a percentage, to show the dynamics of changes.

Summarizing the values collected from analyses for the 3 research objects, it should be noted that for each of them a change in the NDWI value was noted by about 20-30% in relation to the initial state, i.e. before the construction of the water reservoir.

Moreover, in each case it was a change that allows to state that the amount of water in plants has increased. However, due to the small number of analyzes for a single object, the far-reaching conclusions should be formulated carefully. In order to confirm or refute the thesis about the correlation between the construction of the reservoir and the increase of the NDWI value, it is necessary to perform analyzes for the possibly long period to calculate the average from many years. As in the case of the Landsat mission satellites there is a regular return time over the same area, it is the analysis of time series. Such analysis will give the final answer to the question about the correlation for the NDWI index. However, the authors believe that this work gives encouraging results for the further research.

4.3 Analysis of changes in the SAVI and TSAVI indices for the selected water reservoirs

As in previous cases, analyzes of the state before and after construction of water reservoirs were made to determine their impact on the value of the index. Analyzes were made as before for three test objects, i.e. Dobczycki, Klimkowski and Domaniowski Reservoirs.

The same test fields as for the NDWI analyses were used. When calculating the SAVI value, the L parameter value was assumed L = 0.5. All results were collected in the appropriate tables (Tab. 5 and 6).

Summarizing the SAVI and TSAVI analyzes, a decrease in the value of indices was observed for all three test objects. These changes ranged from approx. 24% to approx. 29% for the first index and from 21 to 43% for TSAVI. As in the case of the NDWI index discussed earlier, before the final conclusions are drawn, analyzes for a longer period should be performed and only on the basis of multi-year results compiled with similar ones obtained for comparative objects (ie for which no water reservoir acted), the final thesis can be formulated.

Table 5. The results of SAVI and TSAVI analyzes for Dobczycki and Klimkowski reservoirs. Source: own research.

Test field	Dobczycki Reservoir						Klimkowski Reservoir					
	SAVI 1984	SAVI 2016	Change %	TSAVI 1984	TSAVI 2016	Change %	SAVI 1993	SAVI 2016	Change %	TSAVI 1993	TSAVI 2016	Change %
1	0.6237	0.4459	-39.88	0.4220	0.3047	-27.80	0.5793	0.4430	-23.53	0.3028	0.3934	29.91
2	0.5861	0.4228	-38.63	0.3974	0.2897	-27.10	0.6403	0.4262	-33.44	0.2919	0.4335	48.49
3	0.3997	0.4133	3.28	0.2752	0.2834	2.98	0.7739	0.5968	-22.88	0.4033	0.5212	29.23
4	0.5741	0.4214	-36.24	0.3895	0.2887	-25.87	0.6728	0.3854	-42.72	0.2653	0.4547	71.42
5	0.5636	0.4988	-13.00	0.3827	0.3393	-11.34	0.7268	0.5404	-25.65	0.3664	0.4901	33.75
6	0.6379	0.4458	-43.07	0.4314	0.3047	-29.38	-	-	-	-	-	-
7	0.6837	0.5265	-29.86	0.4615	0.3574	-22.56	-	-	-	-	-	-
average	**0.5813**	**0.4535**	**-28.20**	**0.3943**	**0.3097**	**-20.15**	**0.6786**	**0.4784**	**-29.64**	**0.3260**	**0.4586**	**42.56**

Table 6. The results of SAVI and TSAVI analyzes for Domaniowski reservoir. Source: own elaboration.

Test field	Domaniowski Reservoir					
	SAVI 1994	SAVI 2016	Change %	TSAVI 1994	TSAVI 2016	Change %
1	0.496	0.452	-8.85	0.340	0.309	-9.01
2	0.391	0.350	-10.28	0.270	0.243	-10.2
3	0.441	0.302	-31.6	0.303	0.211	-30.42
4	0.445	0.299	-32.94	0.306	0.209	-31.71
5	0.363	0.284	-21.72	0.252	0.199	-20.88
6	0.599	0.380	-36.61	0.407	0.262	-35.74
average	**0.456**	**0.344**	**-23.67**	**0.313**	**0.239**	**-22.99**

5 Conclusions

In the conducted research, changes in the structure of land use surrounding the given water reservoir were analyzed along with an attempt to determine the dynamics of these processes. On the basis of analyzes performed using the K - means method, it was found that as a result of the construction of a water reservoir, at a distance of 500 m from the shoreline the percentages of built-up land, forests and wasteland in the total area of the terrain increase while the percentage share of agricultural land and pastures decreases.

For each of the analyzed objects, the trend was similar. The authors, however, believe that the impact of socio-economic changes on changes in land use should also be taken into account.

The subsequent analyzes focused on the analysis of changes in the NDWI (Normalized Difference Water Index), which determines the water content in plants. For this indicator, the most promising results were obtained. It was found that due to the construction of a water reservoir, the NDWI value increases from about 20 to 30%, which means that the proportion of water in plants increases.

On the basis of the analysis of changes in the SAVI (Soil Adjusted Vegetation Index) and TSAVI (Transformed Soil Adjusted Vegetation Index) indices, it was found that these indices, correlated with the state of vegetation, as a result of the construction of water reservoirs decreased respectively from 20 to 30% in the case of SAVI and from 20 to nearly 40 in the case of TSAVI.

Summarizing, it can be assumed that the above mentioned indices are useful for modelling the impact of construction of water reservoirs on the surrounding environment. On the basis of satellite data, the percentage impact of water reservoirs on the change in NDWI, SAVI, TSAVI values or land use change was determined. In order to obtain more accurate results, the authors postulate an increase in the number of observations and analysis of results from many years, both before and after the construction of water reservoirs.

References

1. Krajowy Zarząd Gospodarki Wodnej, Diagnosis of the current state of water management (In Polish: Diagnoza aktualnego stanu gospodarki wodnej). Warszawa: Krajowy Zarząd Gospodarki Wodnej, (2010)
2. B. Werner, European waters - current status and future challenges. Copenhagen: European Enviroment Agency, (2012)
3. M. Bodziony, Computer aided design of retention systems delaying flood outflow from the mountain catchment (In Polish: Komputerowe wspomaganie projektowania systemów retencji opóźniających odpływ powodziowy ze zlewni górskiej). Politechnika Krakowska, (2005)
4. A. Byczkowski, *Hydrology* (In Polish: Hydrologia), Vol. **I** and **II**. Warszawa: Wydawnictwo SGGW, (1992)
5. A. Ciołkosz, A. Kęsik, *Satellite remote sensing* (In Polish: Teledetekcja satelitarna). Warszawa: PWN (1989)
6. M. Szostak, P. Wężyk, P. Hawryło Spatio-temporal land use and land cover changes of the Błędów Desert (In Polish: Przestrzenno-czasowe zmiany pokrycia i użytkowania terenu Pustyni Błędowskiej). Problemy leśnictwa w górach i

regionach przemysłowych, Wydawnictwo Uniwersytetu Rolniczego w Krakowie, Kraków, pp. 103-113 (2015)

7. M. Szostak, P. Wężyk, P. Hawryło, M. Pietrzykowski, The analysis of spatial and temporal changes of land cover and land use in the reclaimed areas with the application of airborne orthophotomaps and LANDSAT images. Geodesy and Cartography, **64**, 1, pp. 75-86 (2015)

8. I. Piech, B. Drożdz, Satellite imagery as a source of information about landscape (In Polish: Obrazy satelitarne jako źródło informacji o krajobrazie), Infrastruktura i Ekologia Terenów Wiejskich, No 3/2010, Polska Akademia Nauk, Oddział w Krakowie, pp. 41–54.

9. J. Niedzielko, S. Lewiński, Land cover change detection using Landsat imagery- comparison of three methods (In Polish: Detekcja zmian pokrycia terenu na zdjęciach satelitarnych Landsat – porównanie trzech metod). Teledetekcja Środowiska, Warszawa (2012). ISSN 0071-8076

10. J. Tou, Pattern Recognition Principles. Londyn: Addison-Wesley Publishing Company, pp. 353-354 (1974)

11. C. Tucker, Monitoring the grassland of the Sahel. International Journal of Remote Sensing, **7**, pp. 1571 – 1581 (1984)

12. A. Richardson, J. Everitt, Using spectral vegetation index to estimate rangeland productivity. Geocarto International, **1**, pp. 63-77 (1992)

13. J. Lyou, D. Yaun, L. Ross, A change detection experiment using Vegetation Indices. Photogrammetric Engineering & Remote Sensing, **64**, 2, pp. 143-150 (1998)

14. P. Ceccato, Detecting vegetation water content using reflectance in the optical domain. Remote Sensing of Enviroment, **77**, pp. 22-33 (2001)

15. L. Shunling, L. Xiaowen, J. Wang, Advanced Remote Sensing. Amsterdam: Elsevier, (2012).

16. Z. Bonczar, R. Kostuch, R. Kwiecień, B. Kwoczyńska, Ł. Popławski, J. Słupek , E. Stanke, A. Szymacha, Impact on the environment of the newly built Domaniów reservoir on the Radomka River (In Polish: Oddziaływanie na środowisko nowo wybudowanego zbiornika wodnego Domaniów na rzece Radomce), monograph from the series Infrastruktura i Ekologia Terenów Wiejskich. PAN Komisja Technicznej Infrastruktury Wsi, Kraków, 4 (2004)

17. G. Bo-Cai , NDWI - a Normalized Difference Water Index for Remote Sensing of Vegetation Liquid Water from space. Journal of Remote Sensing of Enviroment, pp. 257-266 (1996)

18. Z. Kurczyński, *Photogrammetry* (In Polish: Fotogrametria). Warszawa: PWN, (2014).

19. Digital Glob (2017). [Online] 07 may 2017. https://digitalglobe.com.

20. N. Nargess Memarsadehi, S. Netanahu, J. Le Moigne, A fast implementation of the ISODATA clustering algorithm. International Journal of Computational Geometry & Applications, pp. 71-103 (2007)

21. J. Rouse, Monitoring the vernal advancment and retrogradation (green wave effect) of natural vegetation. Austin: Remote Sensing Center, Texas, (1973)

22. O. Lillesaeter, Spectral reflectance of partly transmitting leaves: Laboratory measurements and mathematical modeling. Remote Sensing of Enviroment, **12**, pp. 247 – 254 (1982)

23. A. Huete, „A soil-adjusted vegetation index (SAVI)". Remote Sensing of Environment, **25**, pp. 259-309 (1988)

24. F. Baret, G. Guyot, TSAVI: a vegetation index which minimizes soil brightness effects on LAI and APAR estimation. Geoscience and Remote Sensing Symposium, IGARSS'89, 12th Canadian Symposium on Remote Sensing. Vancouver : IEEE Xplore, pp. 1355-1358 (1989)

Terrestial laser scaning in forest invetory based on OptD optimization method

Joanna Janicka[1,*], *Wioleta* Błaszczak-Bąk[2], and *Krystian* Pawłowski[1]

[1] University of Warmia and Mazury in Olsztyn, Faculty of Geodesy, Geospatial and Civil Engineering, Oczapowskiego 2, 10-719 Olsztyn, Poland

Abstract. Nowadays the time of acquiring spatial data is one of the most important parameters during measurements. Laser scanning technology makes it possible to record a huge number of measurement data in a relatively short time. Thus it is possible to acquire and update spatial data on large surface areas in a significantly shorter time compared to classical measurement methods. Terrestrial laser scanning can be used in forest inventory because measurements can acquire millimeter-level of detail. Thus measurement are performed quickly, automatically and provides periodical estimates of many important forest inventory attributes.

1 Introduction

Information about forest resources is collected for various purposes. A management and planning policy for these resources is conducted in forest areas. Depending on the purpose, size of the area and scale of the project, measurements are made in a different range. Measurements are made using different techniques and the required accuracy. The basic method of forest stand measurement is the classical method using tachometry. Unfortunately, this technique that is time-consuming and quite expensive. Of course, it provides high accuracy of measurement, but the task execution time is incomparable to the use of terrestrial laser scanning. Terrestrial laser scanning (TLS) automatically measures the surrounding three-dimensional (3D) space, using millions of 3D points [1].

The main advantage of using TLS in forest inventory is the possibility of fast, automatic and detailed forest documentation. The first work related to the estimation of tree attributes in forest inventory was recorded around 2000, e.g. [2-7]. The motivation to use TLS in forest inventory was to increase work efficiency on sample charts, i.e. to replace "manually" performed tree attribute measurements with those that were automatically taken from TLS data [8 -11].

Forest measurements using TLS can also be a useful tool for forest monitoring, i.e. periodic measurements on the basis of which the changes taking place over time can be assessed.

By referencing an object that has been subject to measurements, terrestrial laser scanning can also be used to update the map for design purposes. A fragment of the measurement results presented in the article presents the scanned area of the historic park belonging to the palace and park complex in Sztynort. The measurements were made in order to update the map for design purposes and as a basis for revitalization works planned for this object. The scope of works included inventory of the forest and a number of analyzes conducted by greenery designers. The measurements were made using the classic method and using TLS. As it turned out, classical measurements took much more time at the field stage, while TLS measurements are carried out very quickly, while their later development is more complicated than in the case of classical data. A collection of measuring points from the so-called scanning "point cloud" is a huge collection of data that also contains a lot of unnecessary data from the environment. Therefore, the authors want to present the possibility of using OptD method of data optimization.

2 OptD method

The OptD is a optimization method described in [12-14] and used to reducing the number of points in the processing of Airborne Laser Scanning point cloud. Of course there is a lot of algorithms described in the literature [15-16] that allows to filter and reduce the point clouds obtained from laser scanning, but the proposed algorithm, in comparison to other methods, allows to obtain the optimal solution. The algorithm of this method can reduce the datasets based on the specified optimization criterion. Such a criterion may be, for example, the percentage of points in the dataset after reduction or the standard deviation. The OptD method can be conducted in two options:

- Option 1: OptD method with one objective optimization (OptD-single).
- Option 2: OptD method with multi objective optimization (OptD-multi).

The most significant stage in OptD method is using the method of generalization. For this propose we used Opheim algorithm, which is often used in the

* Corresponding author: joannasuwm@gmail.com

generalization of curves with complex shapes [17]. The points of a generalized object are examined by means of a tolerance figure built of a circle with a radius of d_{min} and a surface with a length of d_{max}. The initial value of the tolerances are defined by the user, the following values are determined in an iterative process.

If OptD-single method is chosen, then a set which strictly is fulfilling one condition is sought. There is also a possible to process using OptD-multi, then the several results will be obtained, among which the best one should be selected.

In this paper the OptD-single method was tested to optimize data from terrestrial laser scanning. For the OptD-single method the percentage of points to be left in the set after the reduction was chosen as the basic criterion. During the operation of the method, the following are selected: width of the measuring strip and tolerance used in Opheim generatlization method. The values of these parameters are calculated without user intervention and changed in the iterative process in such a way that the optimization criterion is met.

3 Forest inventory with the OptD method

Forest inventory performed by the classical method is a labor-intensive and time-consuming process. Before starting measurements, a geodetic network must be established or the existing one should be found. Unfortunately, GPS measurements cannot always be used to improve the measurements work because in the forested areas it is technically not possible. Then each tree is measured one by one. During measurements, it is important to mark the measured trees not to duplicate measurements or not to miss some.

Using terrestrial laser scanning may be helpful. TLS can be used to evaluate the tree attribute estimations, such as:

- Stem detection and stem density,
- DBH measurements,
- Tree height measurements
- Stem curve, volume and quality measurements
- Biomass estimation.

Thus, you can perform a series of analyses on the acquired data. The results depend on many factors, such as forest structures, the instrument used and processing methods. However, in any case of performed tests and analyses, it is important to pre-filtering the point cloud obtained as a result of the scanner measurement.

The proposed OptD method allows to significantly reduce the volume of data that is processed to obtain the final product, without losing their quality.

4 Example

Maps for design purposes are cartographic studies include the preparation of geodetic documentation necessary for the implementation of an investment project in Poland. The scope and mode of map execution are regulated by various legal acts which define in detail the final result of the entire process. Usually surveying works related to the map update task are associated with measurement in the field using classic measurement technologies such as tachometry or GNSS satellite measurements.

Updating the map for design purposes in the area of property located in Sztynort consisted in measuring trees: determining their location and determining selected features, such as height, diameter, etc. The area of the park is about 40 ha. The classic method ensures high accuracy of measurements, however, it is extremely time-consuming. The use of GPS technology was not possible because the crowns of trees obscured the visibility of the satellites. Therefore, an attempt was made to use also a technique of terrestrial laser scanning, in addition to the classical measurement technology.

As it was mentioned in the introduction a collection of measuring points from the scanning, "cloud of points", is a huge collection of data that also contains a lot of unnecessary data from the environment. Therefore, the authors want to present the possibility of using OptD methods of data set optimization.

The set of measurement points from scanning, "point cloud", is a huge data collection, which also contains a lot of unnecessary data from the environment. Therefore, the authors want to present the possibility of using OptD methods to optimize the data set.

The amount of data obtained as a result of scanning from forestry area (a lot of details, e.g. leaves, branches, shrubs) is very large, but incorrect reduction or filtration may result in the loss of valuable information. Therefore, it is important to optimize the data sets in such a way that with the minimum of data we retain maximum information.

The analyzes were carried out in three stages, successively for the forest area, single tree and trunk. The aims of performed tests were (a) to present how the reduction carried out on the entire (initial) dataset influences the set of points representing a single tree, and (b) does the tree after reduction retain its geometry, and the points representing the tree can be used for 3D modeling. This is particularly important for green area designers, biomass calculation, state and age of the forest.

In the first stage a forest fragment was used for the tests. This fragment was measured from three scanner stations, so the 3D image was composed of three scans. Selected forest fragment before processed the data with the use of OptD method included about seven hundred thousand points. Separation of individual tree trunks, and then modeling was quite complicated, because there were many observations that disturbed the picture. Thus, at the initial stage of the work, the collection was reduced using the OptD method. Optimization of the dataset at various levels was performed. The obtained results are presented in figure 1.

 https://doi.org/10.1051/e3sconf/20186300009

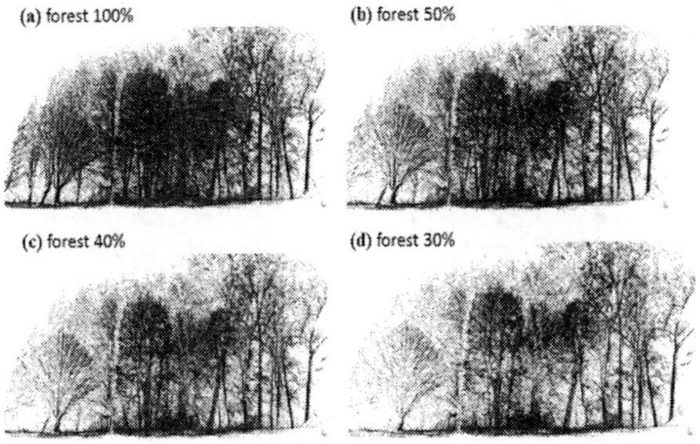

Fig. 1. Dataset without optimization with OptD method (forest 100%) and after optimization wit OptD: a), b) and c).

The characteristics of obteined dataset after OptD method are presenetd in table 1.

The following parameters were used to characterize the reduced datasets: *Zmin, Zmax, Zmean,* standard deviation (SD). The OptD method allowed keeping *Zmin* and *Zmax* values, the average value of the height in the set will change and the SD parameter means the range of the height of the measurement points in relation to the mean. SD will increase as the number of points in the point cloud decreases.

Table 1. Characteristics of obtained parameters for the forest.

	Parameters				
Dataset	Z_{min} [m]	Z_{max} [m]	Z_{mean} [m]	**Number of points**	**SD [m]**
forest 100%	-3.029	34.713	9.895	682 467	7.683
Forest 50%	-3.029	34.713	10.410	341 608	8.831
Forest 40%	-3.029	34.713	10.512	271 489	9.219
Forest 30%	-3.029	34.713	10.642	202 861	9.658

The optimization was carried out in a few scenarios, leaving 50, 40 and 30% of the starting set. The maximum reduction number of points was reduced to 30%, so approximately 200,000 points. Unfortunately, the cloud of points, which still contains a huge amount of data, does not show clearly its features in a small drawing. The numerical data presented in the table 1 show some values, which confirm that even a large reduction of data does not worsen the quality of the resultant dataset.

The next stage of the study was to test whether the optimization using the OptD method allows to significant reduce the size of the data for a single tree but without losing the quality of the data. The number of points representing this single tree presented on the figure is more than twenty thousand points. And in the case of a single tree, you can reduce the size of the data set to 30% of the original size and still the full validity of the tested object is assured.

The figure 2 shows a single tree based on the original dataset and datasets after OptD method. The shape of the tree has been preserved, despite the use of various degrees of reduction.

The characteristics of obtained dataset after OptD method are presented in the table 2. The number of points has been reduced from 22,000 to less than 7,000, and the standard deviation has changed slightly.

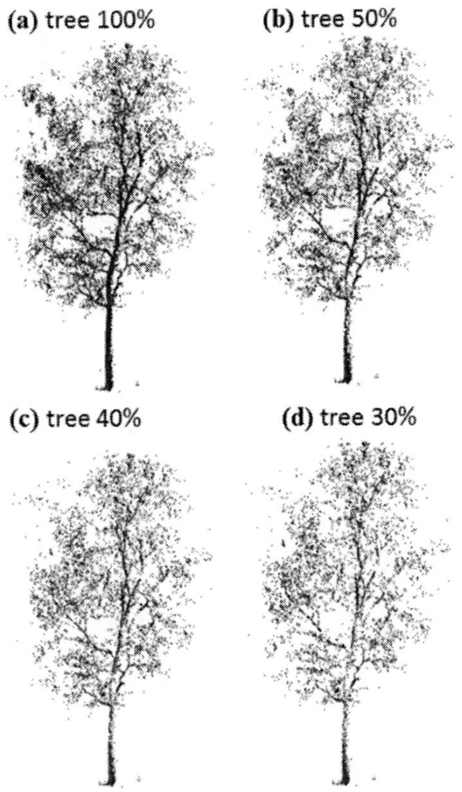

Fig. 2. Dataset without optimization with OptD method (tree 100%) and after optimization wit OptD: a), b) and c).

Table 2. Characteristics of obtained parameters for a single tree.

	Parameters				
Dataset	Z_{min} [m]	Z_{max} [m]	Z_{mean} [m]	**Number of points**	**SD [m]**
Tree 100%	-1.094	31.157	13.557	22 008	7.458
Tree 50%	-1.094	31.157	15.457	10 982	7.276
Tree 40%	-1.094	31.157	15.673	8 886	7.431
Tree 30%	-1.094	31.157	15.919	6 649	7.706

The last stage of the study consisted in reducing the number of points representing the tree trunk. The authors wanted to show that after data reduction, even at the level of 70 from the original dataset, the diameter, perimeter or height of the trunk can still be accurately determined.

The results are presented in figure 3.

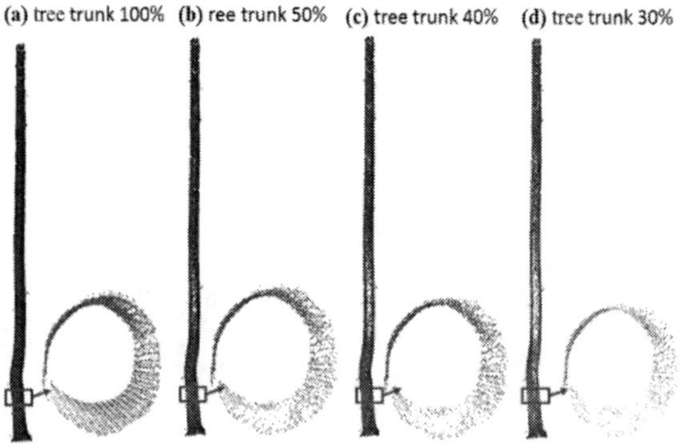

Fig. 3. Dataset without optimization with OptD method (forest 100%) and after optimization wit OptD: a), b) and c).

As it was mentioned in the description of the method, it is important to preserve the three-dimensional features of the object. The results show that the cross-section of the tree trunk after the reduction of points to the level of 30% maintains the correct shape. The trunk perimeter calculated on the basis of the original data tree (trunk 100%) at the height of 1.3 m is 1.36 m. The same perimeter value was obtained for all reduced sets (trunk50%, trunk40%, trunk30%).

The characteristics of obtained dataset for tree trunk after OptD method are presented in the table 3.

Table 3. Characteristics of obtained parameters for tree trunk.

	Parameters				
Dataset	Z_{min} [m]	Z_{max} [m]	Z_{mean} [m]	**Number of points**	**SD [m]**
trunk 100%	-1.645	14.763	3.922	71 158	4.057
trunk 50%	-1.645	14.763	3.958	35 409	4.573
Trunk 40%	-1.645	14.763	4.004	28 503	4.774
Trunk 30%	-1.645	14.763	4.071	21 507	5.024

As in previous cases, extreme values of heights have been preserved, and the value of the SD parameter for the tree trunk has increased by 0.967 m.

5. Summary

The aim of the study was to show that terrestrial laser scanning can be a valuable tool when performing forest inventory and study of single tree. Performing a scan of the area that is covered by the study is a quite simple task. Scanning takes a relatively short time, but at the also very important step is to develop the acquired data. In this paper OptD-single method was used in preliminary processing of TLS data. Performed analysis confirm the efficiency and effectiveness of the OptD method in the process of developing data from forest inventory. When significantly reducing the number of measured points, the remaining points represent the object with the appropriate accuracy. Therefore, any analysis can be performed without fear that the optimized dataset lost their quality. The details conclusions are: (a) OptD method can be used for entire dataset representing forest, but analysis can be performed for single object (tree, tree trunk), (b) the single tree after reduction preserved correct shape, (c) the same value of tree circuit was obtained for all reduced datasets.

References

1. X. Liang, V. Kankare, J. Hyyppä, Y. Wang, A. Kukko, H. Haggrén, X. Yu, H. Kaartine, A. Jaakkola, F. Guan, M. Holopainen, M. Vastaranta, Journal of Photogrametry and Remote Sensing, **115**, Pages 63-77 (2016)
2. M. Erikson, V. Karin, *Proceedings of the Scandlaser Scientific Workshop on Airborne Laser Scanning of Forest*, pp. 244–250 (2003)
3. Lovell et al., Can. J. Remote Sens. **29** pp. 607-622 (2003) DOI:10.5589/m03-026
4. Hopkinson et al., Can. J. For. Res., **34** (2004), pp. 573-583, DOI:10.1139/x03-225
5. G.G. Parker, D.J. Harding, M.L. Berger, J. Appl. Ecol., **41** pp. 755-767 (2004) DOI:10.1111/j.0021-8901.2004.00925.x
6. M. Thies, N. Pfeifer, D. Winterhalder, B.G.H. Scand., J. For. Res., **19,** pp. 571-581 (2004)
7. Henning and Radtke, Photogramm. Eng. Remote Sens., **72** p. 1349 (2006)
8. A. Bienert, S. Scheller, E. Keane, G. Mullooly, F. Int. Arch. Photogramm., Remote Sens. Spatial Inform. Sci., **36** (2006)
9. H.G. Maas, A. Bienert, S. Scheller, E. Keane, (2008). Int. J. Remote Sens., **29,** pp. 1579-1593, Doi.10.1080/01431160701736406
10. M. Vastaranta, T. Melkas, M. Holopainen, H. Kaartinen, J. Hyyppä, H. Hyyppä, (2009). Photogramm. J. Finland, **21**, pp. 51-61.
11. G.E. Murphy, M.A. Acuna, I. Dumbrel, (2010). Can. J. For. Res., **40**, pp. 2223-2233 (2010) DOI:10.1139/X10-171

12. W. Błaszczak-Bąk, Acta Geodynamica et Geomaterialia. **13,** 4(184), pp. 379–386 (2016) DOI:10.13168/AGG.2016.0020.
13. W. Błaszczak-Bąk, A. Sobieraj-Żłobińska, M. Kowalik, Measurement Science of Technology, **8**, 7, 075009 (10pp). (2017) DOI:10.1088/1361-6501/aa7444
14. K. Bakuła , *Challenges of Modern Technology, WUT PhD Students Board*, **2**, No. 4, pp. 31-36 (2011)
15. B. Bauer-Marschallinger, D. Sabel, W. Wagner, Computers & Geosciences, **72,** (2014) DOI:10.1016/j.cageo.2014.07.005
16. W. Błaszczak-Bąk, A. Sobieraj-Żłobińska, B. Wieczorek, *E3S Web Conf. 26 00005*, (2018), DOI: 10.1051/e3sconf/20182600005
17. H. Opheim, Smoothing a digitized curve by data reduction methods. Eurographics '81, ed. by Encarnacao, J. L., pp. 127–135 (1981)

https://doi.org/10.1051/e3sconf/20186300010

Reconstruction of an architectural object. Case study: Citadel fort No. 33 "Krakus" in Krakow

Izabela Piech[1,*], *Bogusława* Kwoczynska[1], and *Artur* Ciszewski[2]

[1] University of Agriculture in Krakow, Faculty of Environmental Engineering and Land Surveying, Poland
[2] Graduate of the University of Agriculture in Krakow, Poland

Abstract. The aim of the study was to recreate, in the form of a 3D model, the Citadel fort No. 33 "Krakus" in Krakow. The data on the basis of which the three-dimensional model was made were obtained using a Leica ScanStation P40 terrestrial laser scanner, which is owned by the Faculty of Environmental Engineering and Geodesy of the University of Agriculture Hugona Kollataj in Krakow. The scope of field work included performing laser measurements, and then processing the point cloud in the Leica Cyclone 3D program and creating a full architectural model in SketchUp 2016.

1 Introduction

Photogrammetry is now used in many areas of life [1]. Thanks to it, we deal with a wide range of methods and devices that enable obtaining precise data about the world around us. Particularly noteworthy are the very popular satellite techniques and laser scanning, which is among others the subject of this study [2,3]. Due to the simplicity of operation and high precision of measurement, laser scanning has become an extremely valuable tool, applicable in many areas of life [4,5,6,7]. One of the areas that has gained completely new opportunities thanks to laser scanning is architecture [8,9,10,11]. The methods and devices used in this technology proved to be extremely useful at the stage related to the design, construction, inventory and reconstruction of the architectural object [12,13,14]. The proof of this is e.g. creating three-dimensional models based on the point cloud obtained using terrestrial laser scanners [15]. Laser scanning is an active system of data acquisition, thanks to which it is possible to precisely determine the shape and mutual geometrical relations between objects in the scope of the scanner [16]. Thanks to this, it is possible to obtain any views, projections and sections based on once recorded data, perform analyzes for any place of the object as such needs arise, easily repeat measurements, compare them and present results [17].

Scanning technology is currently one of the fastest growing technologies on the market. Many years have passed since the introduction of the first laser scanner and this technology has found its application in various industries. The interest in this subject is still growing. Technological development means that 3D scanners are systematically improved, which directly affects the extension of areas in which they can be used: from surveying and industrial surveys, through medicine, forestry, forensics and many others [15]. Laser scanning allows the measurement of both giant terrain by measuring measurements from the terrestrial orbit (Satellite Laser Scanning - SLS) or using scanners mounted in aircraft (Aviation Laser Scanning - ALS) through smaller linear objects (Mobile Laser Scanning - English. MLS) or point (Terrestrial Laser Scanning - called TLS) to quite small, several-dozen-centimeter objects. Data obtained in the results of laser measurements can be used for various types of analyzes, records, inventory. This technology is applicable not only in geodesy, but also in many related areas such as real estate management and spatial planning. It is also used in architecture (inventory of monuments), civil engineering or even in the investigation of car accidents (examination of car body deformations).

2 Research work

The aim of the study was an attempt to recreate, in the form of a 3D model, the Citadel fort No. 33 "Krakus" in Krakow. It was built as part of the popular then, derived from the German school, polygonal system of building forts. It essentially consisted in applying in the defense premises of a caponier and straight or broken walls. However, it must be noted that the Krakus fort has an eclectic form, therefore, instead of a fairly general term "citadel", it should be described as a transitional fort, between a redoubted fort and a centered one. The first one was characterized by a reduit, that is, the object of the circumference intended for the artillery placed on the foreground of the defensive line, while the centered fort had all the battle stations combined with each other. (Fig.1).

Fig. 1. Krak's mound and ruins of the fort - today's state.

* Corresponding author: rmpiech@cyf-kr.edu.pl

 https://doi.org/10.1051/e3sconf/20186300010

The Krakus Fort is different because instead of the classical reduit, it possessed the so-called "neck barracks" at the entrance to the facility (Fig.2).

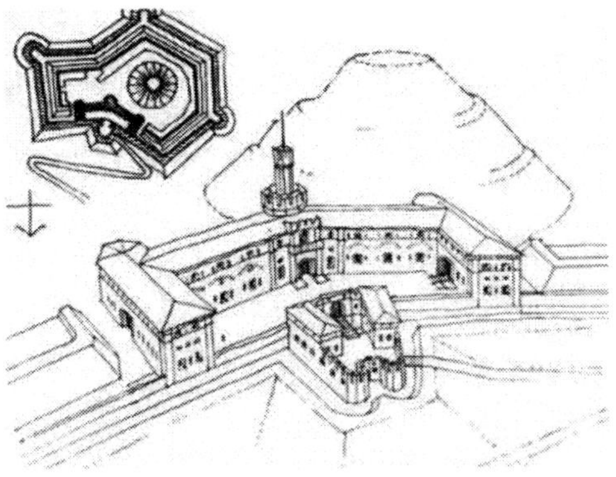

Fig. 2. Neck barracks used in Krakus fort.

Built in the neo-Gothic style, it had the shape of an irregular octagon composed of two connected figures: hexagon around the mound and adjacent quadrangle (Fig.3). There were three caponiers in the defensive circuit reinforced by the embankment, while in the neck of the establishment the neck caponier and already mentioned two-storey U-shaped neck barracks. They were topped with a characteristic two-part tower, reminiscent of old English castles [3]. In terms of building material, fort no. 33 was no different from other forts in Krakow. The basic materials were wood, earth and brick and stone wall [3,18].

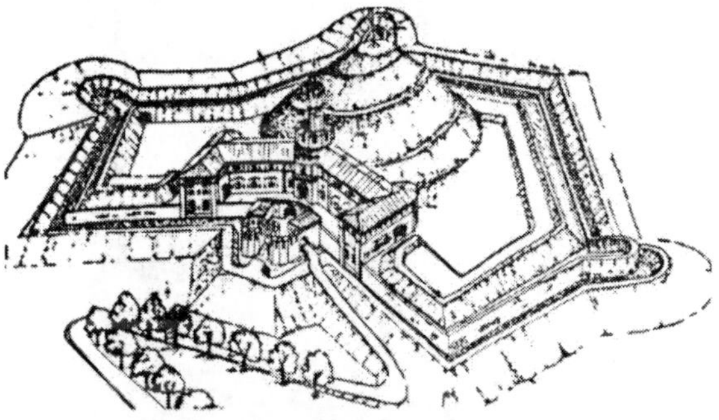

Fig. 3. Visualization of the Krakus fort.

The measurement of the mound and the area around it was made with the Leica ScanStation P40. Field work took place in very good conditions of sunny and almost windless weather. It was decided to scan from six points so that the area of mutual coverage was provided. In addition, 18 reference points in the form of frosted glass spheres were evenly distributed over the measured area, necessary in the process of linking scans from individual positions. The number of measuring points and spheres were dictated by the nature of the measured area and the need to obtain as few dead zones as possible, with simultaneous registration of the reconstructed relics of the fort wall. The scanning was carried out at a resolution of 0.05 m per 100 m both vertically and horizontally. Despite the good layout of field works, the appearance of dead zones, which are mainly located in the place of the barracks, could not be avoided. The end result of field work was obtaining six images of point clouds along with 18 images of reference spheres (Fig.4).

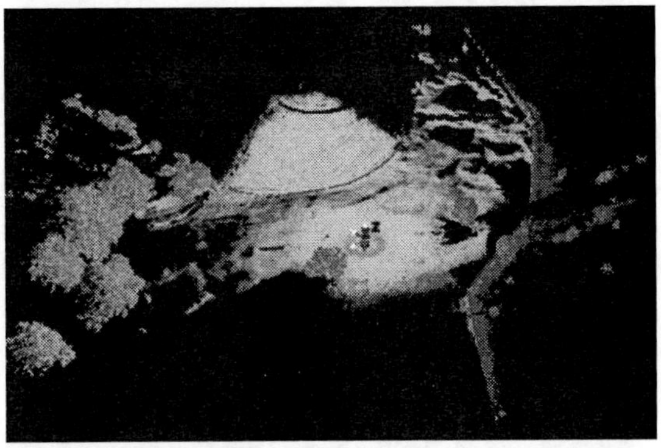

Fig. 4. An image of a single point cloud.

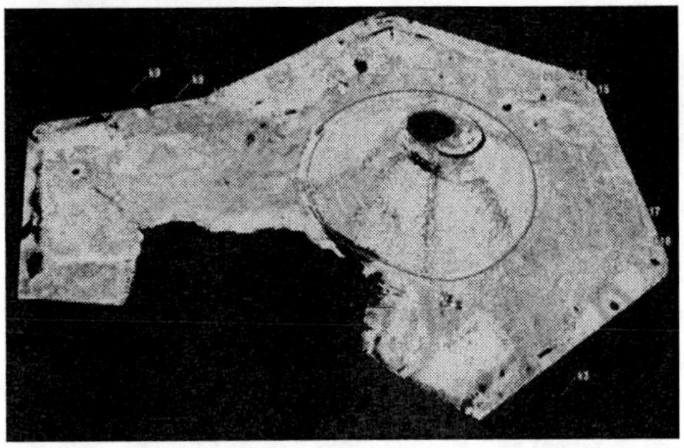

Fig.5. Point cloud after cleaning.

The point cloud post-processing was carried out in the program Leica Cyclone 3D. The first stage of the work was the orientation of scans. The orientation was performed with the accuracy of 6 mm, however, it was necessary to exclude from the calculations three connections of clouds where the accuracy exceeded 2 cm. The result of the orientation was the creation of a point cloud with the measurement noise, which in the further elaboration had to be removed (Fig. 5).

The next stage of the work on the point cloud was the generation of a meshgrid to obtain the terrain model being the basis for further modeling (Fig. 6).

Then modeling of the fort body was made to create a three-dimensional object at the LoD2 level, therefore, the focus was on the main elements of the fort (barracks, retaining wall, earth embankments). The dimensions of individual components of the Krakus fort were adapted to the surrounding area, using the analogy to the Kosciuszko Fort preserved to this day. The work began with modeling the wall surrounding the whole complex, because its relics have been reconstructed and could be captured during the field measurement, and what follows - they were on the terrain model. This allowed for a

https://doi.org/10.1051/e3sconf/20186300010

fairly precise determination of their course. The whole defensive perimeter was raised to a height of approx. 6 m above ground level, so that it constitutes 2/5 of the height of the highest point of the fort, that is the Krakus Mound (Fig.7).

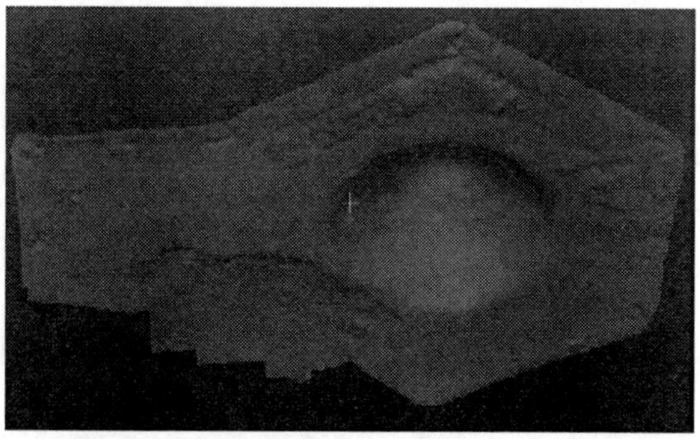

Fig. 6. Mesh generated on a point cloud.

Fig. 7. Retaining wall of the Krakus fort.

The next modeled elements were the neck barracks together with the neck caponier. This fragment was so problematic because it could not be measured during the field measurement. The lack of data from this place was partially completed in the process of the *mesh* generation, because the CloudCompare program automatically gave this terrain the base height, elevated a few meters above the terrain model. In addition, this process allowed to determine the approximate range of the barracks in the place where the *mesh* height was changed. For this reason, the barracks were placed on the internal side of the wall on the whole surface of the height, on the other side, a caponier lying symmetrically between the arms of the garrison building was modeled. Such location is confirmed by archival drawings. In addition to the main body, the buildings were also enriched with a tower standing in the central part of the barracks, door and window openings and access to the entire complex (Fig.8).

The last elements of the fort model were earth embankments, the range of which could not be determined in the field. However, their course was determined in an approximate way thanks to the modeling of the retaining wall and on the basis of available sources. It was decided to place them in parallel to the wall at a distance of about 4 m and finish in front of the barracks, maintaining a distance of 10 m. The front of the embankment, similarly to the archival drawings, is located slightly above the whole embankment. In addition, the embankment has been supplemented with an entry ramp located directly behind the barracks. The model has been subjected to texturing. This process involves the presentation of object surfaces using textures, that is, bitmap or procedural images [19,20]. It was decided to do the texturing manually. In the case of a retaining wall and the barracks, the classic building materials of Austrian fortresses in the form of brick textures and roof tiles were used, while in the case of the earth embankment, the clayey-gold color available in the palette of colors was used (Fig.9).

The last stage of the development of the object was to equip it with elements available in the Google 3D Warehouse gallery. Two types of gates placed in the caponier and in the barracks building as well as the grill in the entrance gate to the entire object were used from the available database, which undoubtedly gives the Krakus fort model the character of a real fortress (Fig.10).

Fig. 8. Barracks and caponier neck during modeling.

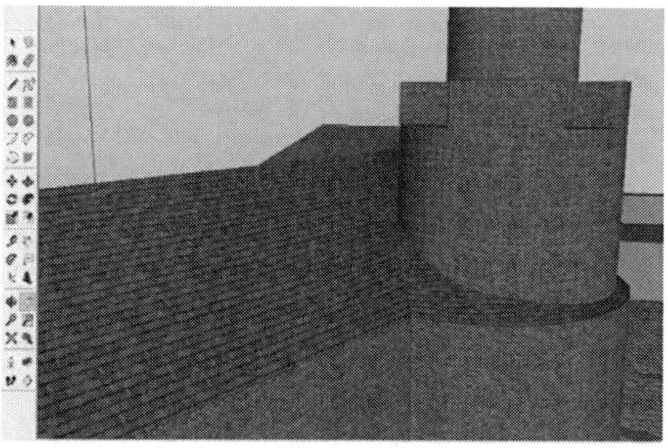

Fig. 9. Model during texturing.

After creating the model, 337 photos based on 7 designed scenes of the object view were exported from the program. The final modeling effect is shown in Figures 11-16.

Fig. 10. Model of barracks gate.

Fig. 11. The barracks of the Krakus fort - view from the north.

Fig. 12. Fort Krakus – isometric projection.

Fig. 13. Fort Krakus - view from the east.

Fig. 14. Fort Krakus - view from the south.

Fig. 15. Fort Krakus - view from the west.

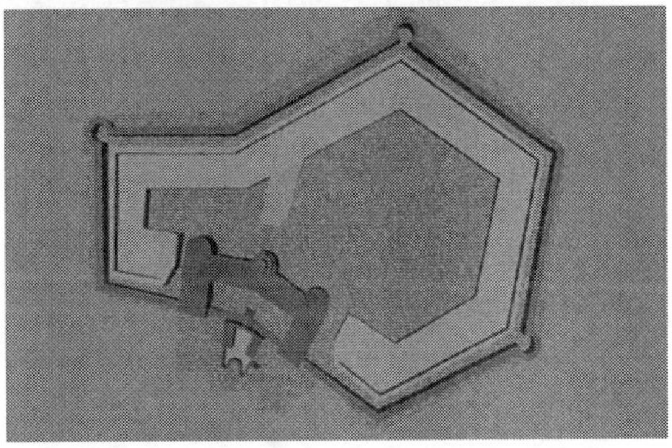

Fig. 16. Fort Krakus - view from above.

3 Conclusions

The desired goal of the work has been reached. The Krakus Mound and the surrounding area where once the fort was located were measured using the terrestrial scanning method. The measurement was made from six stations located around the mound with a resolution of 0.05 m per 100 m. The entire surveyed area was scanned apart from the barracks site, which had a significant impact on the further course of action. The obtained data were developed in the program Leica Cyclone 3D. The edition of the point clouds included a number of activities: orientation of point clouds, clearing the composite image, unification and creation of the *mesh*. The orientation was made with the accuracy of 6 mm. After unification, the grid of triangles was generated on the point cloud, which resulted in obtaining the terrain model. This model became the basis for further development of the architectural object – the fort modeling. Due to the small amount of archival materials, the model of the fort was created at the level of detail of LoD2. For this reason, the focus was on its most important elements, namely the retaining walls, the neck barracks with the caponier and the earth embankment. After modeling the individual elements, the surface of the blocks was covered with textures and the model was enriched with equipment elements. As a result of the performed activities, the geometrical model of the Krakus fort was obtained. Due to the field difficulties and poor archival sources, creation of the model was difficult and probably it is not a faithful reflection of how the fort looked like in reality. However, this does not change the fact that the final effect gives a good image of its appearance and can be re-used.

References

1. Z. Kurczynski, R.Preuss, *Basics of photogrammetry*, Oficyna Wydawnicza Politechniki Warszawskiej, Warszawa, pp. 5-9 (2000) (in polish)
2. A.Ciszewski, *Elaboration of an architectural object using photogrammetric methods*, Masters work. Faculty of Environmental Engineering and Land Surveying, University of Agriculture in Krakow (2018)
3. K. Becek, P. Gawronek, P. Klapa, B. Kwoczynska, P. Matula, S. Mikrut, B. Mitka, I. Piech, M. Makuch *Modeling and visualization of 3D data based on photogrammetric measurements and laser scanning*, Monography (2015)
4. S. Al-kheder, Y. Al-shawabkeh, N. Haala, *Developing a documentation system for desert palaces in Jordan using 3D laser scanning and digital photogrammetry*, Journal of Archaeological Science, **36**, 2, pp. 537-546 (2009)
5. M. Bernat, A. Janowski, S. Rzepa, A. Sobieraj, J. Szulwic, *Studies on the use of terrestrial laser scanning in the maintenance of buildings belonging to the cultural heritage*, 14th SGEM GeoConference on Informatics, Geoinformatics and Remote Sensing (2014)
6. Čučaković, M. Dragović, M. Pejić, M. Srećković, J. Pandžić, *The possibilities of application of 3d digital models in cultural heritage objects' protection and reconstruction*, ICGG Conference monGeometrija, Proceedings Belgrade (2016)
7. E. Friedli, P.W. Theiler, *Documentation of Remote Archaeological Sites – A Comparison Between Long-Range Laser Scanning and UAV-Photogrammetry*, FIG Congress Engaging the Challenges – Enhancing the Relevance Kuala Lumpur, Malaysia 16 – 21 June (2014)
8. H.K. Dhonju, W. Xiao, V. Sarhosis, J.P. Mills, S. Wilkinson, Z. Wang, L. Thapa, U.S. Panday, *Feasibility study of low-cost image-based heritage documentation in Nepal, The International Archives of the Photogrammetry*, Remote Sensing and Spatial Information Sciences, **XLII-2/W3**, 3D Virtual Reconstruction and Visualization of Complex Architectures, Nafplio, Greece, pp.237-242 (2017)
9. M. Doneusa, Ch. Brieseb, M. Feraa, U. Fornwagnera, M. Griebla, M. Jannera, M. Zingerlea, *Documentation and analysis of archaeological sites using aerial reconnaissance and airborne laser scanning*, XXI International CIPA Symposium, 01-06 October, Athens, Greece (2007)
10. Th. Kersten, M. Lindstaedt, K. Mechelke, B. Vogt, *Terrestrial Laser Scanning for the Documentation of Archaeological Objects and Sites on Easter Island*, Computer Applications and Quantitative Methods in Archeology - CAA'2010 Fco (2010)
11. J.L. Lerma, A.E. Seguí, M. Cabrelles, N. Haddad, S. Navarro, T. Akasheh, *Integration of laser scanning and imagery for photorealistic 3D architectural documentation*, C.-C. Wang (Ed.) Laser Scanning, Theory and Applications, INTECH Open Access Publisher, pp. 413–430 (2011)
12. J.L. Lerma, S. Navarro, M. Cabrelles, A.E. Seguí, N. Haddad, T. Akasheh, *Integration of Laser Scanning and Imagery for Photorealistic 3D Architectural Documentation*, Published by World's largest Science, Technology & Medicine Open Access book publisher, pp.413-430 (2011)
13. T.Temizer, G. Nemli, E. Ekizce, A. Ekizce, S. Demir, B. Bayram, F.H. Askin, A.V. Cobanoglu, H.F. Yilmaz, *3D documentation of a historical monument using terrestrial laser scanning case study: Byzantine water cistern, Istanbul,* International Archives of the Photogrammetry, Remote Sensing and Spatial Information Sciences, Volume **XL-5/W2**, XXIV International CIPA Symposium, Strasbourg, France (2013)
14. B. Mitka, A. Rzonca, *Integration of photogrammetric and 3D laser scanning data as a flexible and effective approach for heritage documentation*, 3rd ISPRS International Workshop 3D-ARCH: 3D Virtual Reconstruction and Visualization of Complex Architectures Trento, Italy, International Archives of Photogrammetry, Remote Sensing and Spatial Information Sciences, **XXXVIII-5/W1** (2009)
15. A.Krooks, S. Kaasalainen, V. Kankare, M. Joensuu, P. Raumonen, M. Kaasalainen, *Predicting tree structure from tree height using terrestrial laser scanning and quantitative structure models,* Silva Fennica, **48,** 2, article id 1125 (Online) The Finnish Society of Forest Science The Finnish Forest Research Institute.
16. M.Kędzierski, A. Fryśkowska, M. Wilińska, *Terrestrial laser scanning of engineering and road objects.* Biuletyn WAT, **LIX**, 2 (2010)
17. B Mitka, *Usability of terrestrial laser scanners for the process of documentation and modeling of*

https://doi.org/10.1051/e3sconf/20186300010

historical objects, Archiwum Fotogrametrii, Kartografii i Teledetekcji, **17b**, pp. 525-534 (2007)

18. J. Bogdanowski, *Fortifications and greenery of the Krakow fortress*, Wydawnictwo Literackie, Kraków, pp.84-85, 90-91, 135, 253, 257 (1979)
19. J. Uchański, P. Falkowski, L. Soerensen, *The problem of standardization in the field of architectural object inventory using terrestrial laser scanning techniques*, Archiwum Fotogrametrii, Kartografii i Teledetekcji, **18**, Międzyzdroje, pp. 633-641 (2008)
20. L.Uchański, L. Soerensen, *Technology of terrestrial laser scanning in the problems of reverse engineering and analysis of dynamic processes*, Archiwum Fotogrametrii, Kartografii i Teledetekcji, **21**, Wrocław, pp. 415-424 (2010)

https://doi.org/10.1051/e3sconf/20186300011

Standard deviation as the optimization criterion in the OptD method and its influence on the generated DTM

Wioleta Błaszczak-Bąk[1], *Anna* Sobieraj-Żłobińska[2,*]

[1]University of Warmia and Mazury in Olsztyn, Faculty of Geodesy, Geospatial and Civil Engineering, Institute of Geodesy, Oczapowski St. 1, Olsztyn, Poland
[2]Gdańsk University of Technology, Faculty of Civil and Environmental Engineering, Department of Geodesy, Narutowicza St. 11/12, Gdańsk, Poland

Abstract. Reduction of the measurement dataset is one of the current issues related to constantly developing technologies that provide large datasets, e.g. laser scanning. It could seems that presence and evolution of processors computer, increase of hard drive capacity etc. is the solution for development of such large datasets. And in fact it is, however, the "lighter" datasets are easier to work with. Additionally, reduced datasets can be exchange/transfer/download faster via internet or cloud stored. Therefore the issue of data reduction algorithms/methods is continuously relevant. In this paper authors presented the results of the study whether the standard deviation of measurement data can be used as optimization criterion in the process of dataset reduction conducted by means of the OptD method. The OptD is based on the cartographic generalization methods. In iterative process irrelevant points are being removed and those that characteristic are being preserved, what in results means more points in complex fragments of scanned object/surface and less in flat/uncomplicated area. Obtained reduced datasets were then the basis for DTMs generation. For DTMs assessment RMSE was calculated.

1 Introduction

DTM (Digital Terrain Model) is one of the most popular model representing the surrounding world. It is used in many fields of science and economy to study the state of existing or to simulate future events [1, 2]. It can be developed, inter alia, on the basis of LiDAR (Light Detection and Ranging), which in a relatively short time provides a large amount of measurement data. However, the measurement data itself is insufficient. Additionally, measurements should include a measure of their reliability or uncertainty of measurement [3]. The uncertainty of measurement data is discussed in the Guide to the expression of uncertainty in measurement developed by the International Organization for Standardization (ISO) in consultation with a number of global scientific and technical organizations [4, 5]. Uncertainty of measurement is expressed e.g.: by measurement error, which represents the difference between the measured value x and the real x_0. The measurement error defined in this way is not a measure of the accuracy of the measurement method, because similar measurements made with another instrument at another time and place will give another values [6]. Thus, x_i is a random value and its scattering is characterized by a parameter called the standard deviation estimator SD.

$$SD = \sqrt{\frac{\sum(x_i - \bar{x})^2}{n-1}} \quad (1)$$

where: x_i - observed values, $\bar{x}$ - mean value of these observations, n - number of observations.

SD can be identified with uncertainty of measurement if any of the values x_i is taken as result of measurement.

An important aspect in the development of LiDAR data for DTM generation is an evaluation of the accuracy of the obtained product [7-9]. The accuracy of the DTM is a function of a number of variables such as roughness of the terrain surface, interpolation function, interpolation methods, and the three attributes (accuracy, density and distribution) of the source data [10, 11]. Therefore, the measurement uncertainty parameter, i.e. SD of source data has an impact of the generated DTM.

In this paper focus is on the influence of the source data, in particular, whether the standard deviation estimator of ALS (Airborne Laser Scanning) data can be used as optimization criterion in dataset reduction and whether using SD has an impact on the generated DTM. Reduction was performed by means of the Optimum Dataset (OptD) method [12], which allows to preserve points representing characteristics elements in reduced dataset [13, 14, 15]. DTM was generated on the basis of the original dataset (after its filtration) as well as from the datasets obtained after processing by the OptD method.

* Corresponding author: annsobie@pg.edu.pl

 https://doi.org/10.1051/e3sconf/20186300011

The study area is a fragment of the national road No. 16, Sielska street in Olsztyn, located in the Warmia-Mazury. The measurement was conducted by Airborne Laser Scanning. ALS was made by Visimind Ltd. For this study, part of this measurement was selected. Laser scanning angle was 60 degrees, with a frequency of 10,000 Hz scanning. Scanning was performed from a helicopter with speed of 50 km /h at an altitude of 70 m.

Within analyzed area control points were set up. They were measured by means of Leica Viva NetRover receiver. Observation time in each measurement point was 12 seconds, weather conditions were satisfactory. Corrections VRS from ASG EUPOS were applied. The result of the measurement was the dataset consisting of 204 points.

2 Data reduction and DTM generation

The dataset obtained from ALS used in this study contains 74785 points. There are buildings, street, trees within this area. The ALS dataset is presented in Fig. 1.

Fig. 1. Top view of LiDAR point cloud (source: CloudCompare v.2.6.0).

The selected fragments were filtered by using 'adaptive TIN model' method [16] in proprietary software. As a result of the filtration, there are two datasets: the dataset of points showing the topography (topographic surface dataset - TSset) and a dataset of points showing the details points. The topographic surface dataset for TSset consists of 37481 points. It is presented in Fig. 2.

The datasets after filtration contain only points representing the terrain. Next, they were processed by means of the OptD method in order to reduce them. The OptD puts an emphasis on fact, that the reduction is conducted in a way, that the information necessary for the proper performance of a task is not lost. In this case, points representing characteristics of terrain are preserved due to further DTMs generation. To achieved that, the dataset is divided into measurements strips and within each strip linear object generalization method is used. In this paper Douglas - Peucker (D-P) method [17] was used in OptD-single. The D-P method is based on tolerance distance. The calculations are done in a vertical plane, what allows to accurately check each elevation. The degree of reduction depends on the value of mentioned parameters. They are automatically changed during processing to meet the criterion or criteria set by user. In results, the optimal datasets are obtained.

As optimization criteria in the OptD method parameters like: the number of points in reduced dataset (M) and the percentage of points to be in the dataset after processing (p%) were used and tested so far [13-17]. In this paper it was decided to use the standard deviation estimator (SD) of ALS data.

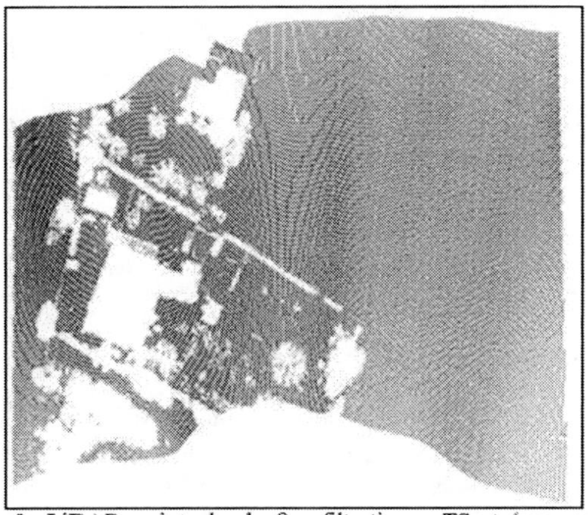

Fig. 2. LiDAR point cloud after filtration - TSset (source: CloudCompare v.2.6.0).

The first calculation was made with assumption, that the difference in SD calculated for original and reduced dataset cannot increase more than SDTSset - SDOptDset <0.200 m. Then the difference of SDs was increased by 0.200 m. In this way, four datasets were determined. The datasets after processing by means of the OptD-single method were named OptDset1, OptDset2, OptDset3, OptDset4 respectively.

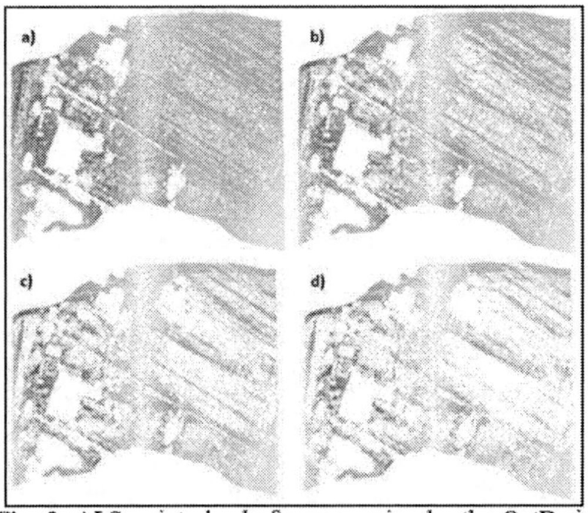

Fig. 3. ALS point cloud after processing by the OptD-single method: a) OptDset1, b) OptDset2, c) OptDset3, d) OptDset4 (source: CloudCompare v.2.6.0).

 https://doi.org/10.1051/e3sconf/20186300011

In order to compare the results of using OptD-single method, the parameters that describe the TSset were determined. Statistics of TSset are presented in Table 1.

Table 1. Statistics of TSset.

Parameter	value [m]
Number of points (N)	37481
Max height [m]	366.361
Min height [m]	348.610
Mean height [m]	358.912
SD[m]	3.760
Point distance	0.340

The OptD-single method selected the optimum solutions for all OptDsets. Statistics of OptDsets, as well as the difference of standard deviation estimator for TSset and for OptDsets are presented in Table 2. For SD OptDset calculation mean height is taken from TSset.

SD calculated for TSset and OptDsets concerns measured height:

$$SD = \sqrt{\frac{\sum(z_i - \bar{z})^2}{n-1}} \quad (2)$$

where: z_i (i = 1,2..., n) – the height of subsequent points in the dataset, $\bar{z}$ – mean height in TSset, n – number of points in dataset (original – N or reduced - M).

The value of SD increases with the decrease in the size of the dataset. In the range from 25% to 70% of reduced points in the TSset, the value of the difference between SDs of TSset and OptDsets ranges from 0.160 m to 0.752 m, these are insignificant values for data representing the surface area.

On the basis of the reduced datasets, DTMs were generated using the kriging method in the Surfer with various grid sizes (0.5m and 1m). In Fig. 4 and Fig. 5 the DTM generated from the TSset (DTM TSset) is shown. Application of the OptD method gave a few solutions, which were used for DTM generation. In Fig. 6 and Fig. 7 four DTMs based on these OptDsets are presented.

To test, how the reduction of measurement data influence the DTMs generation, some cross-sections in random localization were performed. It was made for DTMs with grid 0.5m and 1m. those differences are from range 0.035m -0.149m. For others cross-section results are similar. Thus, it can be assumed, that the reduction did not distorted the generated surface.

In order to determine the accuracy of generated DTM OptDsets within tested area 204 control points were measured by means of GNSS technique. Those points were used to calculate RMSE of generated models.

Table 2. Statistics of OptDsets.

Parameter	OptDset 1	OptDset 2	OptDset 3	OptDset 4
Mean height in OptDset [m]	358.799	358.652	358.480	358.458
Mean point distance in OptDset [m]	0.397	0.491	0.632	0.744
Number of terrain points in OptDset (M)	41488	28955	19610	15136
SD OptDset [m]	3.920	4.121	4.313	4.512
The absolute value of differences of SD TSset and SD OptDset [m]	0.160	0.361	0.553	0.752

Fig. 4. DTM TSset, grid size=0.5m (source: own study in Surfer v.9).

Fig. 5. DTM TSset, grid size=1m (source: own study in Surfer v.9).

 https://doi.org/10.1051/e3sconf/20186300011

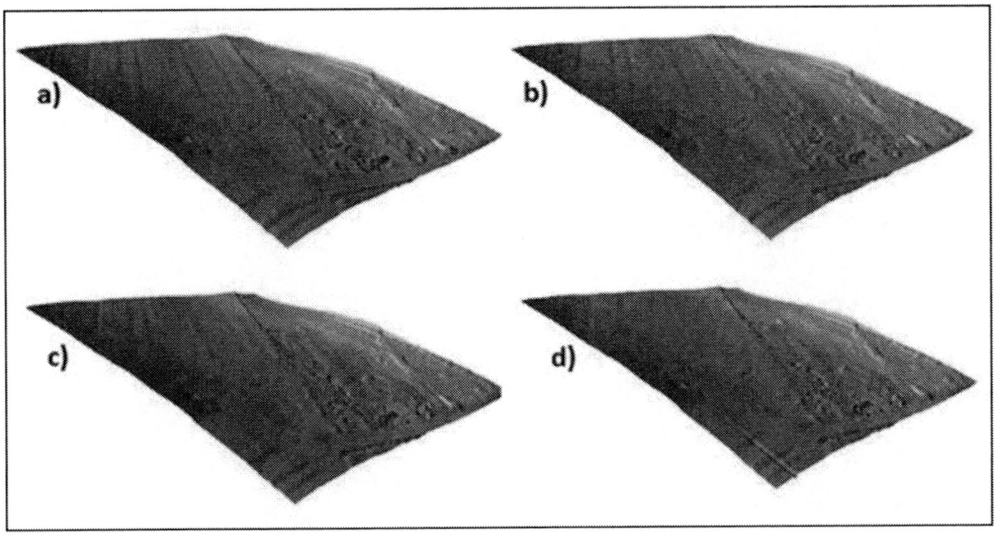

Fig. 6. DTMs based on OptDsets, grid size = 0.5m: a) DTM1, b) DTM2, c) DTM3, d) DTM4 (source: own study in Surfer v.9).

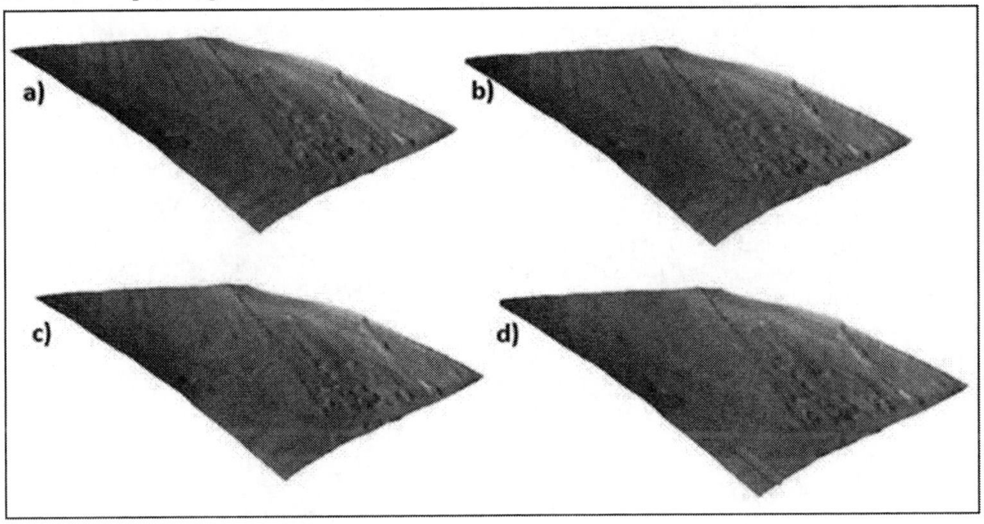

Fig. 7. DTMs based on OptDsets, grid size = 1m: a) DTM1, b) DTM2, c) DTM3, d) DTM4 (source: own study in Surfer v.9).

Table 3: Characteristics of DTM.

	grid size 0.5m				
	DTM TSset	**DTM1**	**DTM2**	**DTM3**	**DTM4**
Zmean [m]	357.988	357.989	357.989	357.990	357.994
RMSE	0.430	0.520	0.605	0.743	0.896
	grid size 1m				
	DTM TSset	**DTM1**	**DTM2**	**DTM3**	**DTM4**
Zmean [m]	357.987	357.988	357.988	357.989	357.992
RMSE	0.510	0.583	0.612	0.782	0.980

Table 4: Differences of RMSE

The absolute value of differences [m]	**DTM TSset – DTMi (i=1, 2, 3, 4)**			
	DTM1	**DTM2**	**DTM3**	**DTM4**
ΔRMSE for DTMs grid size 0.5m	0.090	0.175	0.313	0.466
ΔRMSE for DTMs grid size 1m	0.073	0.102	0.272	0.470

For the generated DTM mean height and RMSE were calculated.

The results show that the values of SD increase along with the decrease in the number of points in the ALS dataset.

Presented results show that the more points have been removed in a dataset by the OptD method, then the bigger are the differences between RMSE of DTMs generated on the basis of this dataset. The range of changes is from 0.073 to 0.470 m. Therefore, using DTMs for various purposes is determined by these values.

3 Conclusions

This paper presents the use of the standard deviation parameter as the optimization criterion in OptD-single method. It was assumed that the difference in SD calculated for the original and reduced dataset cannot increase more than 0.200 m. Tests were carried out for 4 cases, each time increasing the difference in SD by 0.200 m. Next, on the basis of obtained (reduced) datasets, DTMs were generated. For the DTM assessment parameter like RMSE was used. To calculate the RMSE the control points were measured by GNSS.

The analyzes show that the SD of the dataset can be an optimization criterion in the OptD-single method. The obtained datasets met the optimization criterion, and the DTM generated on their basis also met the accuracy expectations. At control points, the calculated RMSE for DTM with 0.5 m grid increased by a maximum of 0.466 m, while for DTMs with 1m grid size by 0.470 m. Differences in RMSE indicated, that application of DTMs generated on the basis of reduced datasets, is determined by degree of reduction.

References

1. K. V. Prasad, *Digital Terrain Modeling: Principles and Methodology*. The Photogrammetric Record, 24: 296–297 (2009) DOI: 10.1111/j.1477-9730.2009.00545_2.x.
2. C. Suchocki, J. Katzer, J. Rapiński, J.,Terrestrial Laser Scanner as a Tool for Assessment of Saturation and Moisture *Movement in Building Materials*. Period Polytech Civ Eng, 1–6. (2018) DOI:10.3311/PPci.11406
3. A. Zięba, P. Ramza, *Standard deviation of the mean of autocorrelated observations estimated with the use of the autocorrelation function estimated from the data, Metrol. & Meas. Syst.*, **18**, pp. 529-542 (2011)
4. ISO/IEC Guide 98-3:2008, *Uncertainty of measurement – Part 3: Guide to the expression of uncertainty in measurement (GUM:1995)*
5. *The Expression of Uncertainty and Confidence in Measurement*, M3003 Edition 3 (2012)
6. S. Bell, *A Beginner's Guide to Uncertainty of Measurement* (1999)
7. J. Estornell, L. A. Ruiz, B. Velázquez-Martí, T. Hermosilla, Analysis of the factors affecting LiDAR DTM accuracy in a steep shrub area, International Journal of Digital Earth, 4:6, 521-538 (2010) DOI: 10.1080/17538947.2010.533201
8. K. Stereńczak, M. Zasada, M. Brach, *The Accuracy Assessment of DTM Generated from LIDAR Data for Forest Area - a Case Study for Scots Pine Stands in Poland*. Baltic Forestry, **19**, 2, pp. 252-262 (2013)
9. Z. Li, Q. Zhu, C. M. Gold, *Digital terrain modeling - principles and methodology*. CRC Press, ISBN 978-0-415-32462-5, pp. I-XVI, 1-323 (2005)
10. Z. L. Li, *Sampling Strategy and Accuracy Assessment for Digital Terrain Modelling*. Ph.D. thesis, The University of Glasgow (1990)
11. Z. L. Li, *Variation of the accuracy of digital terrain models with sampling interval*. Photogrammetric Record, **14**, 79, 113–128 pp. (1992)
12. W. Błaszczak-Bąk, *New Optimum Dataset method in LiDAR processing*. Acta Geodynamica et Geomaterialia, **13**, 4(184), pp. 381-388 (2016) DOI: 10.13168/AGG.2016.0020
13. W. Błaszczak-Bąk, A, Sobieraj-Żłobińska, B. Wieczorek, *The Optimum Dataset method – examples of applications*, ES5 Web of Conferences, **26**, 00005 (2018)
14. W. Błaszczak-Bąk, M. Poniewiera, A. Sobieraj-Żłobińska. M. Kowalik, *Reduction of measurement data before Digital Terrain Model generation vs. DTM generalisation*, Survey Review (2018) DOI: 10.1080/00396265.2018.1474685
15. W. Błaszczak-Bąk, A. Sobieraj-Żłobińska. M. Kowalik, *The OptD-multi method in LiDAR processing*. Measurement Science and Technology. **8**, 7, 075009 (2017) DOI:10.1088/1361-6501/aa7444
16. P. Axelsson, *DEM generation from laser scanner data using adaptive TIN models*, International Archives of Photogrammetry and Remote Sensing, Volume XXXIII, Part B4, Amsterdam, Netherlands:111–118 (2000)
17. D. H. Douglas, T. K. Peucker, *Algorithms for the reduction of the number of points required to represent a digitized line or its caricature*. Canadian Cartographer, **10**, 2, pp. 112-122 (1973)

The concept of anti-collision system of autonomous surface vehicle

Andrzej Stateczny[1], and *Waldemar* Gierski[1,*]

[1]Gdansk University of Technology, Faculty of Civil and Environmental Engineering, Narutowicza St. 11/12, Gdansk, Poland

Abstract. The use of unmanned vehicles in various fields of science and the economy is becoming more common. An extremely important aspect of creating this type of solution is to provide autonomous vehicle navigation, which does not require interference of the human factor or in which it is limited to a minimum. This article discusses the concept of autonomous anti-collision system of unmanned surface vehicle. It proposed a sensor system for the comprehensive implementation of the vehicle path in an automatic manner and presented the idea of remote sensing data fusion.

1 Introduction

The use of unmanned vehicles (UVs) in various fields of science and life is becoming more and more common. Although previously used mainly for military and cosmic purposes, the willingness to study inaccessible places of the earth, gather information about disasters and occurrences at areas of immediate danger to people and facilitate research missions, these experiences were transferred to other applications. The advantages of autonomous vehicles allow their use in scientific research and in the amenities of everyday life. The use of unmanned vehicles for exploratory purposes allows us to work using the smaller human resources involved in measuring service, which makes it more flexible. This saves time and reduces the human factor that can significantly affect the quality of measurements. UVs are characterized by smaller dimensions and weight, which makes it possible to carry out more detailed tests (possible measurement in places where the vehicle on which the people are located could not perform them). Research UVs, as well as those used in everyday life (e.g. for passenger transport) perform their tasks more accurately, their efficiency and attention level does not deteriorate as a result of long working time and fatigue. An additional advantage is low energy consumption, and thus low labor costs. An extremely important aspect of using autonomous UVs for different tasks is detecting obstacles standing in the way of the vehicle in real time. One of the most difficult elements of the concept of an unmanned vehicle is designing the anti-collision system and developing a method of data fusion from various sensors detecting obstacles on a programmed path in order to obtain their reliable avoidance for the safe exploitation of the unit.

The construction of a surface vehicle and development of a sensor system concept requires predicting the conditions under which it will be operated and assumptions in recognizing objects and modelling the environment in which the autonomous surface vehicle (ASV) will move. ASVs are used in many scientific and commercial realizations. They are used by the army for reconnaissance and combat purposes, they serve as research units providing information on various aspects of the aquatic environment or carriers of measuring equipment for the inventory of watercourses and reservoirs. Most of the current solutions work on the basis of remote control. The operator, using the control module, realizes movement of the unit from a certain distance, based on a visual assessment of the surrounding. However, current trends in the development of this type of solutions seek to marginalize human participation in the implementation of the unmanned vehicle path. The main problems in the development of the autonomous navigation system are specific working conditions (contact of two environments with completely different physical characteristics: water – air), difficulty in imitating human senses in navigation with appropriate algorithms or inference structures, as well as in the case of coastal areas, the boundary between water and land. Although the selection of sensors available on the market gives a wide possibility of detecting objects on ASV path, prosaic factors such as the dimensions and weight of the unit are a significant limitation. This restricts the system's recording module, therefore it is time-consuming to develop a design that will ensure optimal detection with the lowest possible usage of batteries. The complexity of environmental conditions and design constraints affects the selection of measuring equipment. The order in which sensors are first selected causes that data fusion and the choice of path implementation methods is an extremely complicated task. All this causes that the development of an autonomous anti-collision system in the case of ASV is a very growing and innovative subject.

* Corresponding author: waldemar.gierski@pg.edu.pl

2 The structure of the anti-collision system

Autonomous anti-collision systems in unmanned vehicles are based on the registration of many sensors. The combination of data from all of them and their proper processing provide an optimal solution for the obstacle-free path of the surface platform. The programmed route (global path) of the vehicle must be supplemented in real time with static obstacles appearing on its way and those in motion relative to unmanned vehicle. In addition, the task of the automatic navigation system is to determine and implement the corrected path. Each of the sensors used to detect obstacles has a number of limitations related to range, detection resolution and optimal working conditions. Proper selection of sensors allows mutual complementation of the obtained information and thus as accurately as possible to determine the parameters of the obstacles necessary to avoid them in the safe way. The structure of the proposed navigation system that can be used in a ASV is given below, taking into account the small dimensions of the vehicle, the conditions of operation and the mutual complementation of data in the navigation process [1,2].

2.1 GNSS receiver and global path planning

Navigation elements using GNSS sensors are the basic assumption of most collision avoidance systems. For proper operation of any sensor it is necessary to determine the ASV exact position in the assumed reference system and to implement detection of objects in its surrounding (in the same reference system). In this case, the GNSS positioning systems are the simplest in application, they give accurate data and enable the simplest interoperability of the anti-collision system with navigational charts and digital terrain models [3]. The GNSS receiver installed on the surface platform enables the implementation of the global path of the vehicle, i.e. the course assumed to be overcome during a specific vehicle mission [4].

Nowadays, GNSS receivers are readily available and there is a huge selection of these sensors in terms of parameters necessary in a given anti-collision system. The one of solution is Trimble R10. The receiver has a small size (11.9 x 13.6 cm) and weight (1.12 kg) when carrying out the measurement will very high accuracy and resolution, sufficient to determine the position in the tracking of the vehicle. It allows measurement with horizontal accuracy: 8 mm + 0.5 ppm RMS and vertical accuracy: 15 mm + 0.5 ppm RMS in real time measurement mode with network transfer of corrections. It has a function of wireless transmission of registration data and is characterized by a low energy consumption of 5.1 W, which argues its use in the discussed system. Tracking satellite signals from more than one satellite system (GPS, GLONASS) affects the reliability of measurement.

Due to the fact that the ASV can be used for measurements in places where terrain apertures may interfere with signal tracking and prevent precise position determination of the antenna, it is proposed to supplement the system. In the case of vehicle operation in a small area (at the shore), the supplement may be a prism system arranged on the deck, whereby it would be possible to determine the position by continuous measurement from the shore of the basin using a total station supported by its own GNSS receiver. In the case of carrying out tasks of the unit in open waters, the solution may be inertial navigation performed by IMU sensors (gyroscope, accelerometer).

An idea of using the IMU sensors is very promising. The second solution of navigation data source is SBG Ekinx-2U. Integrated motion unit with GPS RTK receiver is especially useful during performing hydrographical tasks. Important parameters for anti-collision tasks like position and speed vector accuracy is comparable with Trimble R10 and exceed navigational conditions.

2.2 RADAR sensor

In the case of ASVs, it is very important to detect coastline and keep away from it at the distance of safe maneuvering of the vehicle. The optimal solution for tracking the surrounding of small unmanned platforms is the automotive Doppler RADAR. The principle of operation is the phenomenon of Doppler shift caused by the movement of objects from which the emitted beam of electromagnetic radiation is reflected. When receiving a beam in RADAR, depending on the type of RADAR, the intensity of the received beam is analyzed, the time of return of the beam, and in Doppler's, its frequency, on the basis of which the speed of the moving object is determined. RADAR is used not only to detect objects and determine their position, but also the direction (getting closer, getting away) and speed of getting closer/ getting away (Fig.1. shows the visualization of measurements using RADAR).

The advantages of using RADAR in detecting obstacles are [5]:

- Long detecting range,
- Good velocity estimates,
- Provides nearly all-weather and broad-area imagery,
- High depth resolution and accuracy.

The limitations are:

- Skewed data from fast turning maneuvers depending on the context,
- Limited small and dynamic target detection capability,
- Susceptible to high waves and water reflectivity.

The proposed device for RADAR tracking objects disturbing the path of the vehicle is UMRR Type 42 mobile RADAR. It is characterized by a range of 1.5 to 340 m, a distance measurement accuracy of 0.25 m and a field of view of horizontally: 100° and vertically: 16°. Reflectors having a radial speed component of typ. abs. > 0.1m/s are detected. Having multi target capability, the sensor may detect many reflectors at a time (up to 256) being within the field of view. Depending on the selected communication interface, the number of reported targets may be limited to 128. Targets are sorted by range and if

https://doi.org/10.1051/e3sconf/20186300012

more than 128 are detected, short range targets are reported first. Dimensions of the sensor are around: 230 x 160 mm, a weight: 1290 g. Energy consumption is relatively low while maintaining high accuracy of measurements and amounts to 10.5 W [6].

Fig. 1. Stationary target GRID visualization from the automotive RADAR sensor called UMRR Type 42 Antenna [6].

Appropriate installation of the sensor on the ASV, where the construction elements do not limit the field of view, gives very high quality navigation information, which makes the RADAR the core of the anti-collision system. The importance of using RADAR in this type of vehicles is evidenced by the widespread use of RADARs on ships. RADAR images are particularly helpful when driving a vehicle in an offshore environment or intensively developed areas of hydrotechnical infrastructure (e.g. in port, roads, wharves areas).

2.3 LiDAR sensor

The use of LiDAR sensors in autonomous navigation (anti-collision systems) of unmanned vehicles is associated with their high accuracy, speed and range. Data from the system give a continuous, spatial representation of the surface area around the vehicle, while allowing to determine the velocity of movement and the path of movement in relation to the characteristic elements of the surrounding. A laser rangefinder, sent through a scanning mirror or fiber optic system, scans the area in a plane transverse to the direction of motion. The laser is pulsed and samples the terrain with great frequency, the energy partially reflected from the terrain surface or the obstacles is picked up by the optical system of the scanner and recorded. Based on the measurement of the return time of the reflected signal also allows you to determine the physical features and structure of the object from which the impulse returned to the device. An accurate representation of the environment is required to avoid obstacles. LiDAR data allows orientation of objects and vehicle in all dimensions (3D) of the adopted coordinate system. It gives greater possibilities of anti-collision system. The use of LiDAR in real-time realizations is a difficult task, but with the use of artificial intelligence elements it allows a very accurate visualization of the situation around the vehicle and is extremely helpful in the navigation of unmanned vehicles.

The data from the LiDAR measurement are in the form of a three-dimensional and 360° cloud of points (Fig. 2) for which the coordinates x, y, z in the assumed coordinate system, intensity and beam number (ring number) have been determined. The x, y, z coordinates represent the position of each point relative to signal source. Intensity represents the strength of the signal returning to the receiver (depends on the reflective properties of the object from which the wave was reflected). The beam number determines in which beam the returned point is located [7]. The main advantages of this method of acquiring data about the surrounding are [5]:

• Good at near range obstacle detection,
• High depth resolution and accuracy.

The limitations are:

• Angular resolution both vertically and horizontally,
• Sensitive to environment and ASV motion.

An exemplary sensor that can be used in ASVs is 3D real-time Velodyne VLP-16 LiDAR. Performs measurements in the range of full angle in the horizontal plane and 30° in the vertical plane. It is characterized by compact dimensions and weight (830 g) and low power consumption which additionally supports the legitimacy of its use on small vehicles. Small size retains high accuracy of the registration of the surrounding (16 channels, 100 m range, up to 600 000 points per second). Fig. 2 shows the visualization of measurements using said LiDAR. When implementing the design of the unit, it should be remember that LiDAR, like RADAR, should be located in such a way that other elements of the platform construction do not interfere with the registration area [8].

Fig. 2. Visualization of the point cloud from the mobile LiDAR sensor Velodyne VLP-16 [8].

2.4 Visual sensors

Visual methods are used to detect obstacles due to the simplicity of their use, relatively high efficiency and low installation costs. Stereo vision-based anti-collision systems detect objects that may interfere with the movement of the vehicle on its implemented path. They enable differentiation of objects due to their shape, color and texture, the type of the object and its localization are determined.

Detecting obstacles on stereo images is complicated. It is not know in advance how many objects will be on the each of the input images, it is also not known where in the image they will appear or what dimensions they will have. The problem is quite difficult to solve, hence the method of using artificial neural networks is often used to decode the images into information for the anti-collision system [7].

The advantages of vision sensors are:

• High lateral and temporal resolution,
• Simplicity and low weight in practical application.

The limitations are:

• Low depth resolution and accuracy,
• Challenge to real-time implementation,
• Susceptible to light and weather conditions.

Visual sensors have been selected for the ability to supplement laser (LiDAR) and radar measurements to increase system reliability. Analysis of the RGB stereo camera recording enables noise isolation and exclusion of electromagnetic and laser reflections that are not relevant to the vehicle path. It allows to increase the range of observation and to indicate to other sensors approximate directions and distances to potential obstacles. Interpretation of visual registrations is based on the geometrical relations of the image (the distance, angle and confidence score from the camera are used to create a new direction). The placement and number of sensors on the platform depends on the technological assumptions of its operation. Cameras should be chosen so as to realize the visualization of the surroundings, especially in the direction of vehicle movement.

The installation of a metric cameras system would be the best solution in this type of application of vision methods. However, the complexity and sensitivity of optical systems, large dimensions and weight, as well as high costs limit their use in unmanned units, where space for assembly is limited, and energy consumption and lightweight design prevent their assembly. The solution is widely used, e.g. in unmanned aerial vehicles, non-metric cameras. After the self-calibration process, based on geometrical relationships, they are good source for obtaining relatively accurate photogrammetric data, sufficient for the navigation process [9].

An example of a camera that meets the assumptions of the anti-collision system is DJI ZENMUSE X5S 4K DJI Inspire 2. It enables image recording up to 60 fps in 1920 x 1080 pixels resolution (30 fps in 3840 x 2160). It is a 20.8-megapixel camera with a 4/3" CMOS sensor. It works in the temperature range: -10 to +40°C. It is characterized by reliable image stabilization in the area of: pitch: -90° to +30°, pan: ±320° [10]. These cameras are commonly used in advanced unmanned measuring aircrafts, so it is ideally suited for a floating platform. The development of the camera deployment concept depends on the intended use of the ASV.

2.5 Data acquisition and processing module

The proposed sensors have proprietary software for analyzing and developing the results of their measurements. The data processing system of the autonomous navigation system of the unmanned platform connects these separate applications together, co-creating a uniform result of the vehicle's environment representation. As the remote sensing sensors proposed are intended for mobile applications, the results of their work are sent to the developed analyzer and combined into one coherent information about the location of obstacles in the way of the vehicle.

Data for each sensor should be processed independently and after processing should be fused to establish common collision situation image. Main task for processing algorithms is to detect moving and fixed target. Shore constructions and other nonmoving targets not need target tracking procedure. Anti-collision algorithms for nonmoving targets provide simply rules – not to hit fixed constructions.

For moving targets anti-collision system needs target tracking algorithms. According to previous research the best way is to use neural network tracking filter. Especially useful appears filter based on General Regression Neural Networks. Patented filter consist of two working parallelly General Regression Neural Networks was presented in previous works [11-14].

After detecting of nonmoving targets and moving target tracking procedure data from different sensors should be fused. An idea of different sensors fusion was presented in [15].

Some authors analyzed the various aspects of unmanned vehicles navigation. Video sensor and LiDAR fusion are described in [16]. In [17], an algorithm using LiDAR and a camera for detecting and tracking surface obstacles using the Kalman filter is presented. Legal aspects of ASV navigation, including anti-collision, are described in [18]. An approach using artificial neural networks (ANN) to solve the ASV anti-collision problem is presented in [19-21], where ANN was used to control the autonomous robot. 3D mobile (3D LiDAR) and GNSS applied to autonomous car navigation was presented in [22].

3 Automatic implementation of the path improved by data from the anti-collision system

For the implementation of the unmanned vehicle path complex algorithms are used which, based on the data registered by the sensors, determine the route on which no collision with another object will occur. For this purpose, analyses of fused data collected by the anti-collision system or intelligent methods are used.

The future of this type of systems is marked by the use of artificial intelligence (AI) methods. AI is used in engineering practice to solve problems in the absence of all information necessary for their analytical solution. Simulation of human brain function, the ability to learn from previous system experience, the ability to make decisions in situation where IT systems can not solve the problem are just a few advantages of using AI in the problem of autonomous navigation of unmanned vehicles. Among the many practical applications of AI, artificial neural networks (ANN) are particularly interesting in the process of automatic implementation of the path by the vehicle. The operation of the autonomous anti-collision system, in which activities are undertaken by ANN, depends on the way the network is taught.

Learning is accomplished by the task of conditions according to which the ANN will be able to solve the set task. The effectiveness of ANN depends on the learning phase, the method of action should be properly determined depending on the input signals, the scales of connections neurons responsible for specific tasks and threshold values of neurons should be properly resolved. Correct defining of the network's assumptions allows for the implementation of a specific purpose.

It is also possible to use intelligent algorithms based on fuzzy logic and evolutionary algorithms. Interesting approach of intelligent path planning using evolutionary strategy was presented in [23-25].

In future authors works artificial intelligence method will be tested to provide optimal maneuvering solution for ASV. The most important idea for autonomous vehicle navigation is own ship safety.

4 Summary

Combined analysis of data acquired using various sensors, in this case, LiDAR, RADAR and video sensors, by sensor fusion of navigation sensors will allow to obtain a synergic effect, useful in the aspect of anti-collision maneuvering of future oriented fully autonomous intelligent unmanned vehicles.

The new idea, proposed in the article, not known in available worldwide bibliography, is multisensory fusion set of lidars, radars and cameras for establish intelligent fully autonomous navigation system for ASV based not only on one source of data like it is provided for now but based on fused multisensory data: from those sensors.

References

1. R. Polvara, S. Sharma, J. Wan, A. Manning, R. Sutton, *Obstacle avoidance approaches for autonomous navigation of unmanned surface vehicles*, The Journal of Navigation, **71** (2017)
2. A. Romano, P. Duranti, *Autonomous unmanned surface vessels for hydrographic measurement and environmental monitoring,* FIG Working Week, Rome, (2012)
3. Y. Aotani, T. Ienaga, N. Machinaka, Y. Sadakuni, R. Yamazaki, Y. Hosoda, R. Sawahashi Y. Kuroda, *Development of autonomous navigation system using 3D map with geometric and semantic information*, Journal of Robotics and Mechatronics, **29**, 4 (2017)
4. X. Meng, H. Wang, B. Liu, *A robust vehicle localization approach based on GNSS/ IMU/ DMI/ LiDAR sensor fusion for autonomous vehicles*, Sensors, **17**, 9 (2017)
5. Z. Liu, Y. Zhang, X. Yu, C. Yuan, *Unmanned surface vehicles: an overview of developments and challenges*, Annual Reviews in Control, **41** (2016)
6. UMRR Type 42 Antenna Data Sheet
7. P. Wei, L. Cagle, T. Reza, J. Ball, J. Gafford, *LiDAR and camera detection fusion in a real-time industrial multi-sensor collision avoidance system*, Electronics **7**, 6 (2018)
8. Velodyne VLP-16 Sample Data
9. C.-H. Tsai, Y.-C. Lin, *An accelerated image matching technique for UAV orthoimage registration*, ISPRS Journal of Photogrammetry and Remote Sensing, **128** (2017)
10. ZENMUSE X5S Specification
11. A. Stateczny, *Neural manoeuvre detection of the tracked target in ARPA systems*, Proceedings CAMS 2001, Glasgow, (2001)
12. A. Stateczny, W. Kazimierski, *Selection of GRNN network parameters for the needs of state vector estimation of manoeuvring target in ARPA devices*, Photonics Applications in Astronomy, Communications, Industry, and High-Energy Physics Experiments IV, Wilga, (2005)
13. W. Kazimierski and A. Stateczny, *Optimization of multiple model neural tracking filter for marine targets*, 13th International Radar Symposium (IRS), Warsaw, (2012)
14. W. Kazimierski, G. Zaniewicz, A. Stateczny, *Verification of multiple model neural tracking filter with ship's radar*, 13th International Radar Symposium (IRS), Warsaw, (2012)
15. W. Kazimierski, A. Stateczny, *Fusion of data from AIS and tracking radar for the needs of ECDIS*, Signal Processing Symposium Jachranka, Poland, (2013)
16. J. Jo, Y. Tsunoda, B. Stantic et al., *A likelihood-based data fusion model for the integration of multiple sensor data: a case study with vision and Lidar sensors*, Robot Intelligence Technology And Applications 4, **447**, (2017)
17. L. Jooho, W. Joohyun W., K. Nakwan, *Obstacle avoidance and target search of an autonomous surface vehicle for 2016 maritime RobotX Challenge*, IEEE OES International Symposium on Underwater Technology (UT), Busan, (2017)
18. J. H. Mei, M. R. Arshad, *COLREGs based navigation of riverine autonomous surface vehicle*, IEEE 6TH International Conference on Underwater System Technology, Malaysia, 2016
19. A. Barton, E. Volna, *control of autonomous robot using neural networks*, Proceedings of the International Conference on Numerical Analysis and Applied Mathematics 2016 (ICNAAM-2016), Rhodes, 2016
20. B. Ko, H.J. Choi, C. Hong et al., *Neural network-based autonomous navigation for a homecare mobile mobot*, 2017 IEEE International Conference On Big Data And Smart Computing (BIGCOMP), Jeju 2017
21. T. Praczyk, *Neural anti-collision system for autonomous surface vehicle*, Neurocomputing, **149** (2015)
22. J. Lil, H. Bao, X. Han et al., *Real-time self-driving car navigation and obstacle avoidance using mobile*

3D laser scanner and GNSS, Multimedia Tools and Applications, **76** (2016)

23. P. Kolendo, R. Smierzchalski,, *Ship evolutionary trajectory planning method with application of polynomial interpolation*, Activities in Navigation: Marine Navigation and Safety of Sea Transportation (2015)
24. L. Kuczkowski, R. Smierzchalski, *Termination functions for evolutionary path planning algorithm* 19th International Conference on Methods and Models in Automation and Robotics (MMAR) Miedzyzdroje, (2014)
25. L. Kuczkowski, R. Smierzchalski, *Comparison of single and multi-population evolutionary algorithm for path planning in navigation situation*, Symposium on Mechatronics Systems, Mechanics and Materials, Jastrzebia Gora, (2013)

Accuracy analysis of 3D model obtained by photogrammetric method on the example of historic room from Wisłoujście Fortress

Tadeusz Widerski[1,*], *Karol* Daliga[1]

[1]Gdansk University of Technology, Faculty of Civil and Environmental Engineering, 80233 Gdansk, Poland

Abstract. The article presents a comparison of obtained models of a test object. Close range photogrammetry was used to obtain 3D models. As test object was used one of the rooms located in Wisłoujście Fortress in Gdańsk, Poland. Different models were obtained by using different distribution and number of reference points. Article contains analysis of differences between coordinates of control points obtained from total station measurements and estimated from different 3D models.

1 Introduction

Obtaining spatial information (in the form of point clouds) for the purpose of geometric archiving or spatial analysis of measured objects is a method increasingly used in engineering practice [1, 2, 3]. In most cases, laser scanners are used to obtain a point cloud [4]. They are measuring distance and angles and in that way, a point cloud is created. However, these types of instruments and software used by them are very expensive. For this reason, the method of laser scanning in some cases does not apply. As an alternative method of acquiring spatial data, in form of point cloud, photogrammetric method is increasingly used. Software development and more advanced cameras enable to obtain satisfactory data in a much simpler way and at more competitive prices [1, 5]. This article presents accuracy analysis of 3D model obtained by photogrammetric method using non-metric digital cameras. The analysis was made based on coordinates of arranged photopoints grid within the measured room. The grid consisted of 134 photopoints was measured by total station in local coordinate system. Leica TPS 1200+ total station was used for the measurements.

2 Development of a 3D model

A room located within the area of Wisłoujście Fortress in Gdańsk was used as test field. The room was characterized by a complex geometry. It was built on plan of a ring segment, ceiling is a cross vault. Corridor separated by an additional wall is also a part of the room. The room is equipped with a brock kitchen hood and a chimney hole that further complicates its geometry (Fig. 1 and Fig. 2).

Fig. 1. Cross section of the room.

Fig. 2. "Outside" view of room's model.

The room's spatial model was made using the non-metric photogrammetry method. For this purpose, two cameras with fixed focus lenses were used: Canon EOS 5D (full frame 12 Mpix sensor) with Sigma 50mm f / 1.4 lens and

* Corresponding author: tadwid@pg.edu.pl

https://doi.org/10.1051/e3sconf/20186300013

Canon EOS 350D (with APS-C 8 Mpix sensor) with Canon 28mm f / 2.8 lens. In total, several hundred photographs of the room were taken, which at further study stage were given a geometrical analysis. The spatial model was generated in Agisoft Photoscan Pro software.

2.1. Points position combinations accepted for development

There are two points types in study: control points used to determine accuracy of obtained model and reference points to fit model into local coordinate system.

Different combinations of reference points placements on walls, floor and ceiling were used for analysis. The combinations used, their subsequent determinations and control points number are shown in Table 1. Prim (') designation means setting where used points created another possible diagonal. As reference was used situation when all measured points were used in model development (the case marked "Max").

Example of point positions on different walls was shown at Fig. 3

Table 1. List of accepted combinations of points distribution and control points number.

Determination of the combinations	Points position, their number and symbol							Number of control points
	Wall				floor	ceiling	hood	
	N	S	E	W	D	G	Ok	
1	1	1	1	1	1	0	0	93
1NSEWD Ok	1	1	1	1	1	0	1	93
1NSEWDG	1	1	1	1	1	1	0	93
1NSEWDG Ok	1	1	1	1	1	1	1	92
2N'S'E'W'D'G'	2	2	2	2	2	2	0	89
2N'S'E'W'D'G' Ok	2	2	2	2	2	2	1	88
2NESWDG	2	2	2	2	2	2	0	90
2NESWDG Ok	2	2	2	2	2	2	1	89
4D	0	0	0	0	4	0	0	94
4E	0	0	4	0	0	0	0	96
4E Ok	0	0	4	0	0	0	1	96
4N	4	0	0	0	0	0	0	94
4N Ok	4	0	0	0	0	0	1	93
4S	0	4	0	0	0	0	0	95
4S Ok	0	4	0	0	0	0	1	94
4W	0	0	0	4	0	0	0	94
4W Ok	0	0	0	4	0	0	1	93

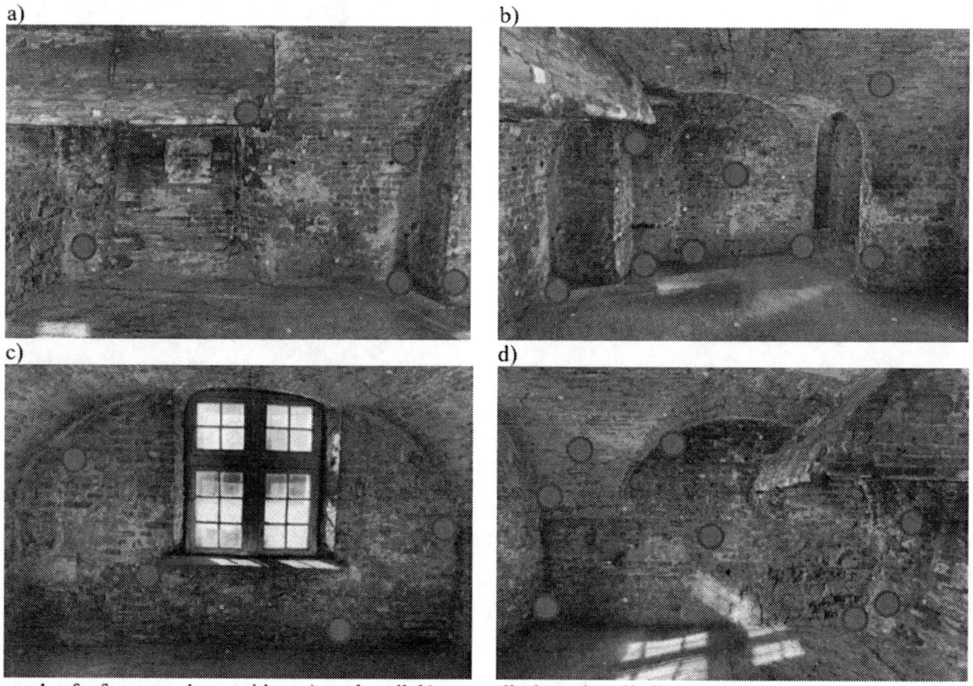

Fig. 3. Example of reference points position: a) north wall, b) east wall, c) south wall, d) west wall.

 https://doi.org/10.1051/e3sconf/20186300013

2.2. Image processing

In the first stage, all measured points on pictures were marked. In addition, measured points coordinates were imported to project. In order to check correctness of point assignment, preliminary merging of photos was performed and it was checked whether any of points generated a large matching error (in this case, level of 1 cm was assumed). In this way, data for the "Max" combination have been prepared. The remaining combinations were created by removing those points that were not supposed to be reference points.

Projects prepared in this way, consisting of photographs with reference points marked on them, have been further processed. Firstly, photos were aligned with "accuracy" parameter set to "high" and with enable pair preselection. Next step was camera's parameters optimization. Next stage, considered in this study, was automatic recognition of points in the pictures, which weren't reference points (control points were created). Deliberately, no points were added manually to check how good results can be obtained from a study done in a maximum automated manner. In this case, used software was able to recognize 96 out of 134 points (71 % of distributed tags). The last step was to export estimated coordinates of control points. In data export, a mathematical coordinate system was used.

3 Measurement results analysis

Based on points coordinates obtained from total station measurement and coordinates estimated from model, differences between coordinates were calculated using the equations:

$$\Delta X = x_{meas} - x_{est} \quad (1)$$

$$\Delta Y = y_{meas} - y_{est} \quad (2)$$

$$\Delta Z = y_{meas} - y_{est} \quad (3)$$

where "meas" index means coordinate from total station measurement and "est" index means coordinate obtained from model.

Additionally, spatial coordinates differences R were calculated using equation:

$$R = (\Delta X^2 + \Delta Y^2 + \Delta Z^2)^{0.5} \quad (4)$$

Calculated differences were used to calculate the first, second (median) and third quartiles. Quartiles were used to compare results due to lack of distribution normality of obtained differences ΔX, ΔY, ΔZ, R. Comparison of obtained quartile values for each combination are shown on Fig. 4 to Fig. 11.

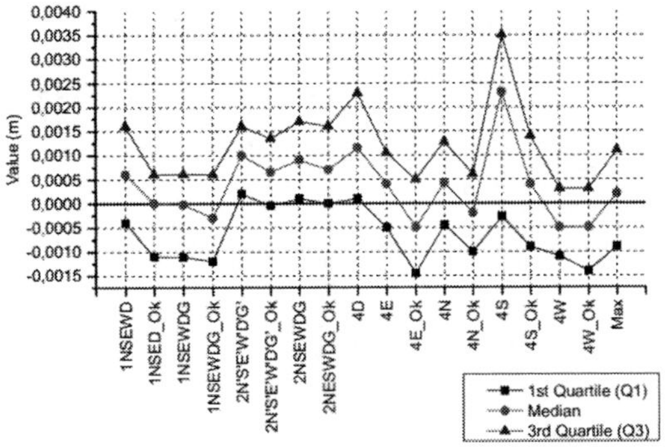

Fig. 4. Quartile values list of coordinate differences for X axis sorted by a reference point combination.

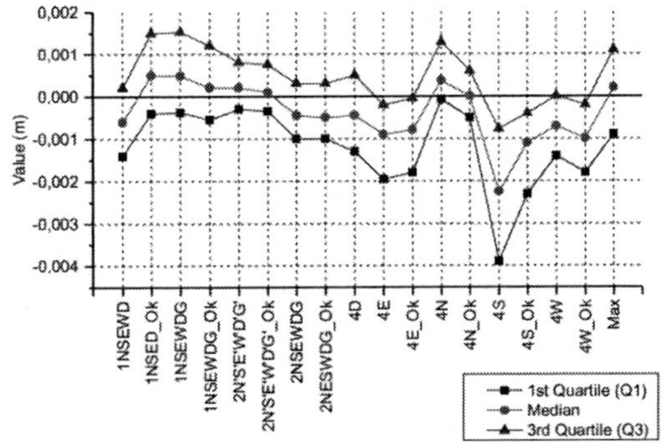

Fig. 5. Quartile values list of coordinate differences for Y axis sorted by a reference point combination.

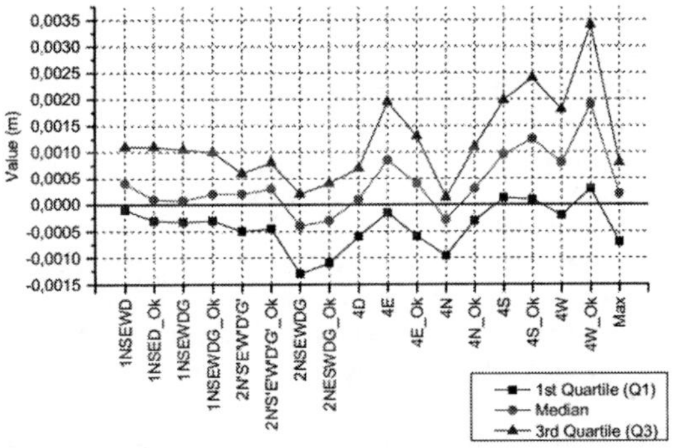

Fig. 6. Quartile values list of coordinate differences for Z axis sorted by a reference point combination.

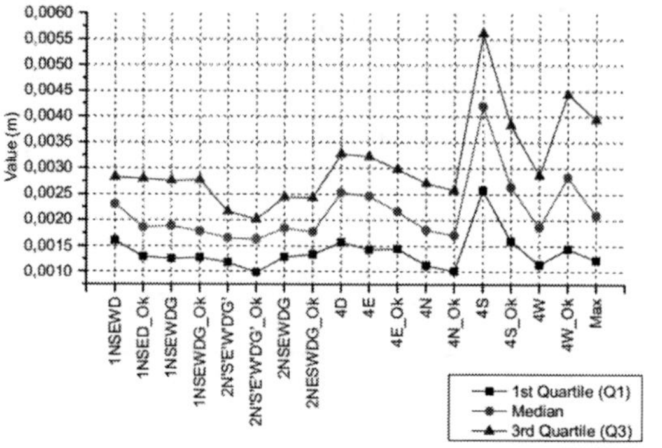

Fig. 7. Quartile values list of "spatial differences" R sorted by a reference point combination.

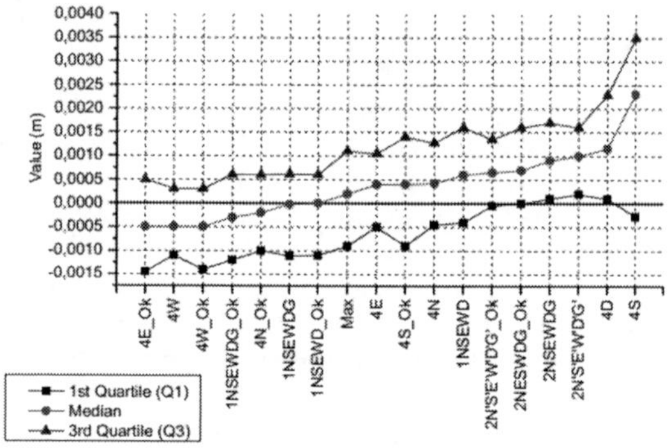

Fig. 8. Quartile values list of coordinate differences for X axis sorted by increasing median value.

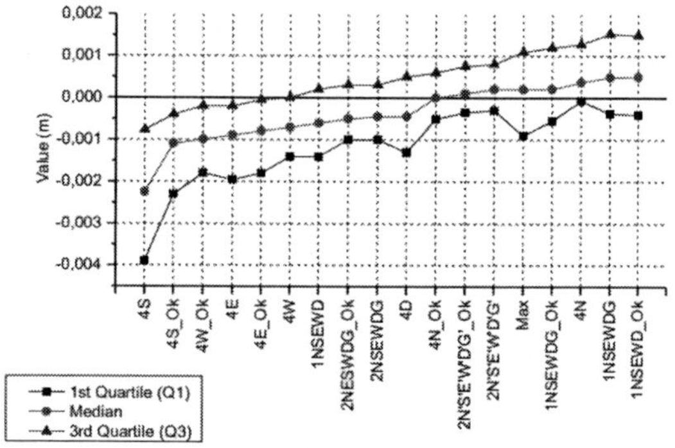

Fig. 9. Quartile values list of coordinate differences for Y axis sorted by increasing median value.

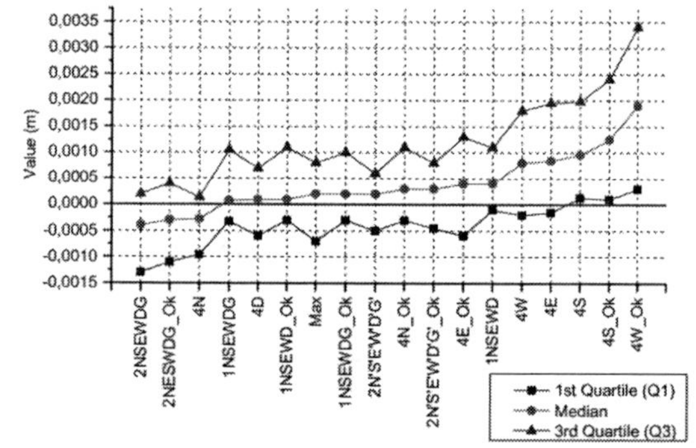

Fig. 10. Quartile values list of coordinate differences for Z axis sorted by increasing median value.

Instrument used for measurements had point position accuracy of an order of 2 mm. For this analysis, it must be remembered that differences less than 2 mm are irrelevant. Analysing data presented on Fig. 4 to Fig. 7, it can be noticed that in most cases, addition of point located on hood to calculations improves correspondence of measured and estimated coordinates of control points. In addition, in the case of a combination related to points located on walls and floors diagonals, differences are practically the same. From considered combinations, the biggest variance between measured estimated values can be seen for combination "4S". In this combination, one of used reference point was located on windowsill and was located horizontally. A significant incidence angle of laser beam caused that point position was measured with a larger error. An error in determining coordinates of this reference point caused a worse model adjustment to coordinate system than when using reference points measured at optimal angle. In case of combinations in which reference points were located only on room's walls or floor, comparable coordinate differences were obtained. In the case of "Max" combination, all reference points were control points at the same time. In this case, it is possible to check used software's ability to reproduce object's geometry. It can be seen that coordinates differences are not greater than coordinates accuracy calculated based on total station measurements. The median for "spatial differences" corresponds to total station's distance measurement accuracy (2 mm). However, the third quartile (75 % of results is less than quartile value) reaches 4 mm, which wasn't observed for other considered combinations. It can be caused by using a larger number of points, some of which were measured at an unfavourable incidence angle of laser beam emitted by total station's rangefinder.

Fig. 8 to Fig. 11 presents charts with sorted median values for various combinations of considered reference points.

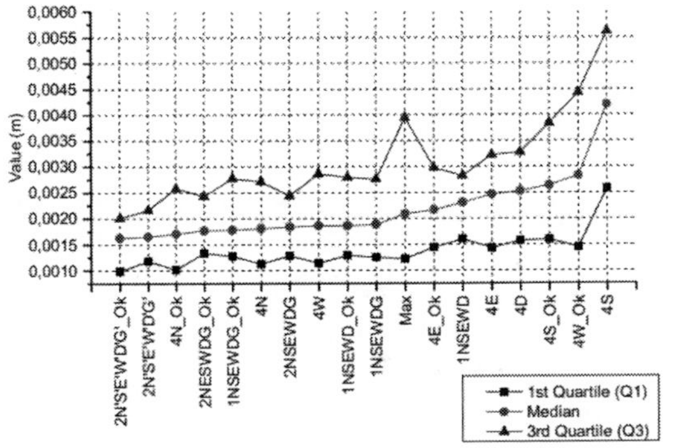

Fig. 11. Quartile values list of "spatial differences" R sorted by increasing median value.

By analyzing the sorted median values for the differences of individual coordinates, it can be noticed that the combinations with the highest number of points gave results comparable to "Max" combination. The worst results were obtained when points were placed only on one wall, but also when one point was placed on all surfaces simultaneously (combination of 1NSEWD). It follows that it is best to measure points that will be in places where object's geometry changes. In this way, time of field measurements can be optimized, because measurements of reference points are one of the more time-consuming processes performed in the field.

4 Summary

In process of making a model based on images, it is important to skilfully arrange markers and take pictures of object. Properly placed and photographed markers enable quick and accurate execution of object's digital model. Reference points should be chosen wisely because their measurement can be one of the longer stages during field work. On the one hand, reducing the number of reference points results in faster measurements. On the other hand, this requires increased attention while measurements so that all measurements will give correct results.

It is important to properly arrange the reference points on measured object. They should be placed in places where a character of object's geometry changes, eg near surface bends. In general, they should "surround" object which will be modelled. In this way, number of points needed to measure will be minimized and obtained results will be in line with expectations.

References

1. A. Barbasiewicz, T. Widerski, K. Daliga, E3S Web of Conferences, *The analysis of the accuracy of spatial models using photogrammetric software: Agisoft photoscan and Pix4D*, 26 (2018), DOI:10.1051/e3sconf/20182600012
2. T. Widerski, K. Daliga, SGEM, *Photogrammetric and total station's measurement of compressed column's model*, 17(22), 427-434, (2017), DOI:10.5593/sgem2017/22/S09.054
3. Z. Ye, X. Tong, Y. Xu, S. Gao, S. Liu, H. Xie, X. Liu, Photogrammetric Engineering and Remote Sensing, *An improved subpixel phase correlation method with application in videogrammetric monitoring of shaking table tests*, 84(9), 579-592, (2018), DOI:10.14358/PERS.84.9.579
4. M. Stankiewicz, J. Reiner, G. Kotnarowski, SPIE - the International Society for Optical Engineering, *Positioning of scanned part inside of the laser triangulation system*, 8082, (2011), DOI:10.1117/12.889469
5. M. A. Soto-Zamora, I. E. Vízcaino-Hernández, R. A. Díaz-Zeledón, J. J. Velasco-González, *Application of digital close-range photogrammetry to the modeling of heritage structural elements for its analysis by FEM*, (2019) DOI:10.1007/978-3-319-99441-3_36

Application of ALS data for GNSS terrain obstacles inventory

Renata Pelc-Mieczkowska[1,*], *Dariusz* Tomaszewski[1], and *Karolina* Jurgielewicz[1]

[1]University of Warmia and Mazury, The Faculty of Geodesy, Geospatial and Civil Engineering, Poland

Abstract. Current constellation of global navigation satellite system (GNSS) ensures signal availability even in severe observational conditions like urban canyon or under tree canopy. However, positioning in such environment remains a challenge because obstacles can block, reflect and diffract GNSS signals which significantly affects accuracy. Those errors are strongly sight dependent and cannot be mitigated in differential positioning that is why, knowledge of the shape and spatial distribution of terrain obstacles is essential. In this paper using of airborne laser scanning (ALS) data for terrain obstacles inventory is presented and evaluated. In proposed method terrain obstacle models have been derived from ASCII ALS data file using open source QGIS with LAStools software suite and dedicated ALSObstModel plugin. Test models were developed for three geodetic control points with different environmental characteristics. For each point reference model from direct tachometry measurements have been obtained. An average error in determining the elevation of the terrain obstacles from ALS based models was 0.6° to 1.7°. This distance corresponds to 3 to 6 minutes of satellite in orbit.

1 Introduction

Wide availability of networks of permanent reference stations and current constellation of global navigation satellite system (GNSS), consists of almost 100 satellites make satellite positioning techniques very popular and readily used. The main limitation of these techniques is the need to provide an access to unobstructed horizon. Some terrain obstacles, like in case of urban canyon [1] or tree canopy [2] can block, reflect and diffract GNSS signals, which significantly affects accuracy. Due to the fact that these errors are strongly sight dependent and cannot be mitigated in differential positioning, knowledge of the shape and spatial distribution of terrain obstacles is essential in proper mission planning as well as GNSS data processing. Regardless of whether we are considering GNSS mission planning or pre-processing GNSS data, we always need to relate the position of the satellites to the terrain obstacle layout. Thus, in the procedure, we separate two steps:

- creation a polar plot of the elevations of the obstacles above horizon
- computation the satellite positions and comparison obstacles' and satellites' elevations [3].

Over the last 20 years, many methods of generating obstacles diagrams have been proposed. These methods can be divided into two categories: first requiring some observations directly at the measuring point, and second one based on spatial data sets. The most obvious method is to use theodolite. This method characterized by the highest accuracy but is very time-consuming and poses some technical difficulties in the case of curtains on high elevations. The more advanced solutions, successfully tested by authors in previous research, are terrain obstacles modelling based on terrestrial laser scanning and digital hemispherical photography [4,5]. The first group of methods allows to obtain very accurate models of terrain obstacles. And it is recommended to use such methods if there is a possibility of quick access to measurement points. In the case of measurement campaigns taking place in large areas, performing direct measurements would consume too much time and resources. Therefore, in such cases, one should look for methods that allows to obtain the appropriate and accurate terrain obstacle models without having to perform a direct measurements.

Therefore, there is a second group of satellite visibility determining methods that includes the most advanced methods based on 3D digital surface models [6] and 3D buildings models. Moreover, 3D buildings models can be used not only for advanced GNSS mission planning, but also for supporting GNSS navigation in urban canyon like in GNSS shadow matching technique [7]. The use of 3D models is, however, burdened with errors in measurement data, model execution errors and the error space model location itself. In addition the main limitation of those methods is the fact, that there are no complete 3D surface models (models including, apart from buildings, trees and other obstacles) for many areas, except large urban agglomerations. Therefore, using 3D models data in forested areas is virtually impossible.

In recent years, LiDAR (light detection and ranging) technology has become a popular method of spatial data obtaining and ALS data become more accessible than ever. For example in POLAND in the years 2011-2014, the LIDAR data for 92% of the country's area was obtained as part of the ISOK project [8]. These data were obtained regardless of the level of urbanization, therefore

* Corresponding author: renata.pelc@uwm.edu.pl

https://doi.org/10.1051/e3sconf/20186300014

it is possible to use them for all points of the geodetic network. From the point of view of terrain obstacle modelling, ALS has a number of advantages, especially provides high dense 3D points cloud in which points represents land surface as well as every objects above it.

2 Object and Source Data

In order to conduct surveys concerning the inventory of terrain obstacles, three GNSS measurement points were selected. The test points were located at the detailed geodetic control network points with catalogue numbers: 0838806, 0839104 and 0838904. For each of the points, catalog coordinates in the flat coordinate system PL-2000 were obtained. All points were located in the campus of the University of Warmia and Mazury in Olsztyn in Kortowo are. When selecting measuring objects, the most important thing was the inventory of various types of terrain obstacles, mainly buildings and trees. The test measurement was carried out on May 23-24, 2018.

Fig. 1. Obstacles above the test points (0838806 top left, 0839104 top right and 0838904 at the bottom).

2.1 ALS Data

In the presented work, data from ALS acquired in 2012 as part of the ISOK was used. which included the area of the study, in the presented work. The obtained data was characterized by the density of points at the level of 12 points per square meter. The data are provided grouped into files, each corresponding to the extent of the sheet of map in the flat rectangular coordinate system "1992" in the scale 1:1250. The area of study was covered by four sheets:

1)N-34-77-D-b-4-3-2-2
2)N-34-77-D-b-4-3-2-4
3)N-34-77-D-b-4-3-4-1
4)N-34-77-D-b-4-3-4-2.

The study uses data in the LAS format. The LAS (LASer File Format) format is a public format for handling and exchanging 3D point clouds between users. It was developed in 2003 by ASPRS - the American Society for Photogrammetry and Remote Sensing, as the LAS 1.0 Format Standard. It is the world standard for handling laser scanning data and has been approved by manufacturers of equipment (scanners) and software designed for processing data from LiDAR.

2.2 Reference Data

For each point reference model from direct tachometry measurements have been obtained. The measurement was made with the TRIMBLE model M3 total station. The device has the ability to perform non-reflective measurements in Direct Reflex technology, which was a very useful option when aiming at high obstacles located near the zenith, where it was impossible to look through the telescope. However, during a very sunny day, the laser spot was hard to see, therefore it is worth to carry out reflectorless measurements in shady places, during cloudy weather or after dark. In addition, a great convenience was the "infinite" sloths, by which the rotation of the telescope and the instrument, when measuring obstacles on subsequent azimuths was very easy.

Fig. 2. Test points (from left respectively: 0838806, 0839104 and 0838904).

As a result of tacheometry measurement, a RAW file with horizontal and vertical angles was obtained. On their basis, the azimuth and elevation of each measured point of terrain obstacle was calculated. Based on these values, a polygon representing so called *open sky* on the sky plot was created.

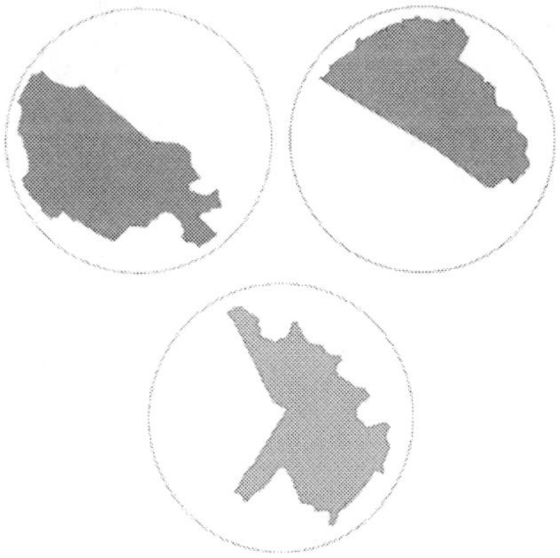

Fig. 3. Reference polygons from tacheometry (0838806 top left, 0839104 top right and 0838904 at the bottom).

 https://doi.org/10.1051/e3sconf/20186300014

3 ALS Data Processing

The main task to do in the presented work was to create terrain obstacle models from ALS data. The input data was four files of point clouds in the LAS format, as mentioned before. For the processing of the ALS data set, a free package created for LiDAR data processing - LAStools was used. This package consists of a dozen or so tools and allows to perform basic operations on point clouds from laser scanning. It was built based on the open source LASlib library (LGPL license) and is available as free for non-commercial purpose. In addition to the LAStools package, another open source software, QGIS, was used to further develop the data. Terrain obstacle model creation was carried out using a dedicated author's QGIS plugin named ALSObstModel.

The obstacle modeling process consisted of several steps. The first of these was to combine all of point clouds covering the area of study into one, using the lasmerge tool of the LAStools package. The resulting cloud was 1 GB in size and was time and sources consuming in further process. Therefore, in the next step, a data set limited by a cylinder was prepared for each considered measuring point. The following assumptions were made during the cutting out of data:

- cylinder axis is parallel to the vertical and passes through a measuring point
- the radius of the cylinder ensures the inclusion of all the objects higher than 10 degrees above the horizon of the GNSS antenna.

Calculation of the cutting radius took place in two stages. At the beginning it was assumed that there are no objects higher than 50 m in the area under consideration and the initial cut off was made. Next, the height of the highest point in the obtained area was determined and on this basis the definitive radius of cutting (d_s) was determined as follows:

$$d_s = \frac{h_{hor}}{\tan 10°} \qquad (1)$$

$$h_{hor} = H_{LiDAR} - H_{St} - h_i \qquad (2)$$

where:
H_{LiDAR} – height of the highest point in cloud, H_{St} – height of measurement point, h_i – instrument height.

Cutting process was performed using lastile tool of LAStools. The obtaining clouds was about 30 KB in size each. Resulting measurement clouds are presented on figure 4.

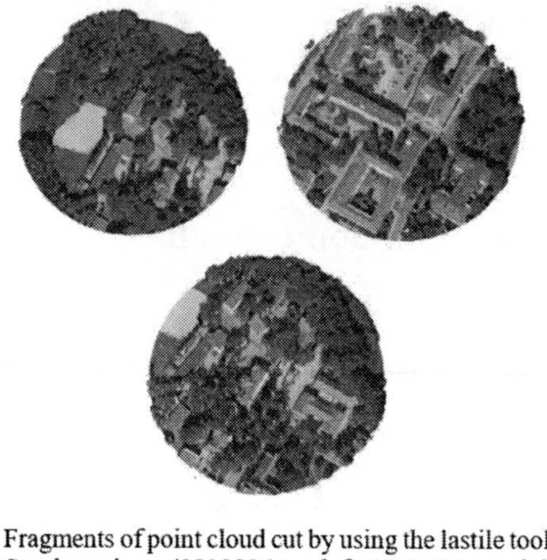

Fig. 4. Fragments of point cloud cut by using the lastile tool from the LAStools package (0838806 top left, 0839104 top right and 0838904 at the bottom).

The ALSObstModel plugin was used to further develop the data. Traditionally, when considering the visibility of satellites, their position is determined by two values: the azimuth and elevation related to the place of observation. The basic function of the ALSObstModel is the conversion of ALS data from The State Geodetic Coordination System 1992 to the azimuth and elevation, taking into account the antenna height.

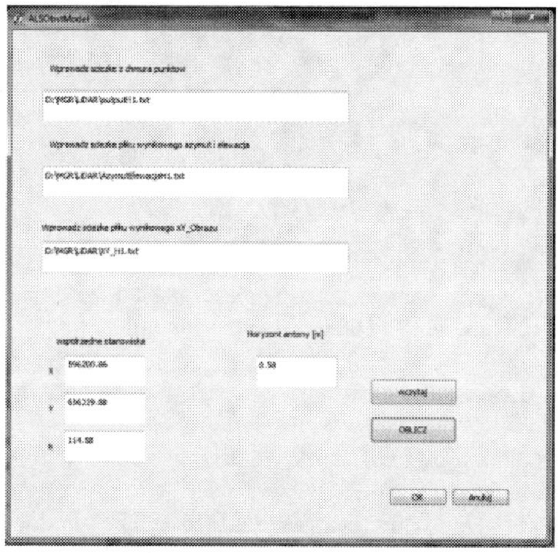

Fig. 5. ALSObstModel plugin window.

The model prepared in this way is saved in the text format and can be used to plan GNSS sessions as well as to pre-process measurement data. Generated files in next step were loaded as CSV text layers into the QGIS application. On their basis, the points were combined into polygons, so that it was possible to compare them with polygons developed from tachometric measurements. The created objects represent an *open sky* as in tacheometry.

Fig. 6. Obstacle models obtained from ALS data (0838806 top left, 0839104 top right and 0838904 at the bottom).

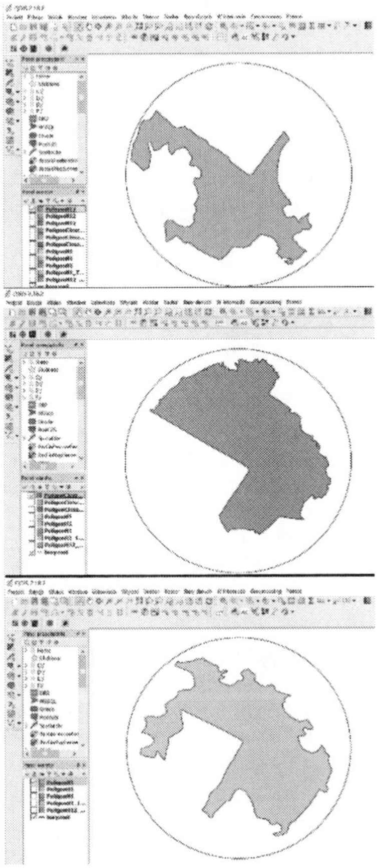

Fig. 7. Polygons representing *open sky,* based on ALS data (from the top respectively 0838806, 0839104 and 0838904).

4. Analysis of Results

From the point of view of determining the GNSS satellites visibility, the measure of the quality of the terrain obstacle model is the accuracy of mapping the boundary between the obstacle and the *open sky*. In order to evaluate the models obtained from the ALS data, these models were compared with the models obtained from direct tacheometric measurement considered as reference. QGIS software was used to compare the obtained models.

The accuracy analysis was performed by comparing the polygons representing the *open sky*. The analysis was performed in the QGIS software. First, using the *geoprocessing/symmetric difference* tool, a polygon layer was created which is the difference of the sum of polygons and their common part.

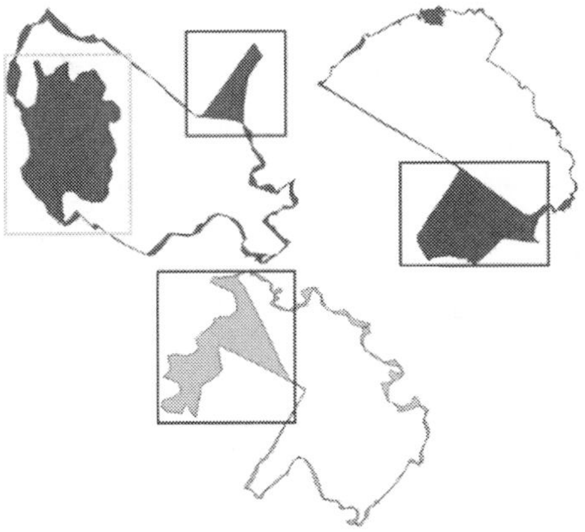

Fig. 8. Polygons represents ALS based terrain obstacle models errors (0838806 top left, 0839104 top right and 0838904 at the bottom).

As it can be seen, in case of each of the models there are areas characterized by errors with large values. Comparing the obtained polygons with hemispherical images of the surroundings of the measurement points, these errors can be classified as gross errors. The source of these errors is twofold. The yellow color indicates the area where errors resulted from the fact that the tree was cut, which grew when the ALS data was acquired. In the red rectangle, however, areas where errors resulted from the non-mapping of the building wall from ALS data were marked.

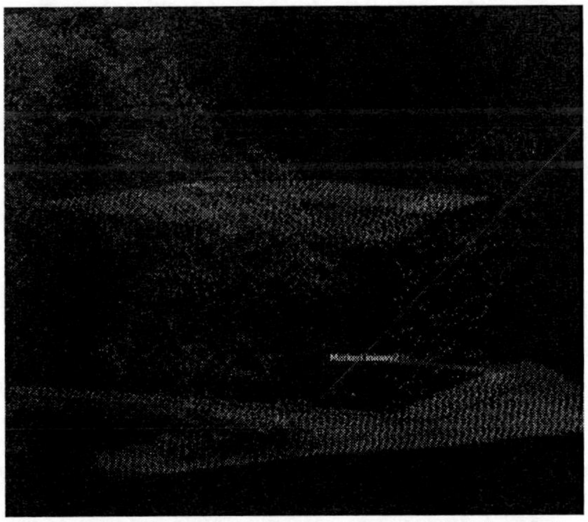

Fig. 9. Small densities of LIDAR points on the building wall.

The lack of building wall on ALS obstacle model results directly from the geometry of the air scanner measurement. As a result of such location of the scanner in the resulting cloud there is a high density of points on the roofs and relatively small density on the walls [9]. The number of wall points depends on the direction and speed of the flight, as well as the scanning density. However, as it can be noticed in the attached figures, regardless of the direction of the flight, densities of the points on the walls are too low to create a suitable model.

In the next step elevation errors for subsequent azimuths (1 degree resolution) were calculated.

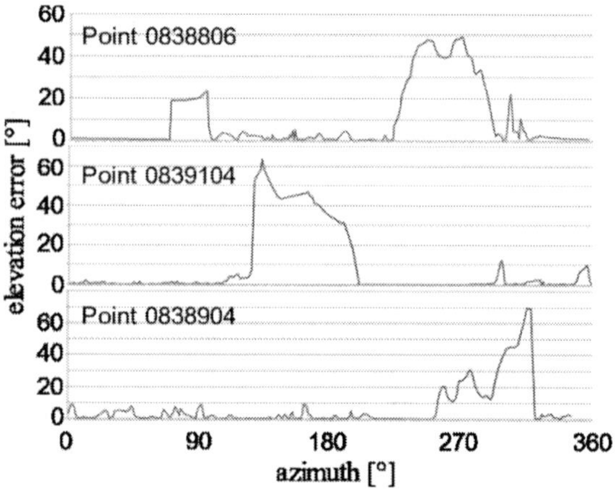

Fig. 10. Terrain obstacle models accuracy.

Taking into account all data appeared errors reaching 60 degrees, marked on the chart in red. However, if we remove data with gross errors from the data set, we will obtain elevation accuracy at the level of single degrees. The maximum error was 22° at point 0838806, 12° at point 0839104 and 9° at point 0838904, while the average errors were 0.6°, 0.7° and 1.7° respectively.

5 Conclusions

The paper presents a method of terrain obstacle modelling based on LIDAR data and analysis of the obtained models. Modelling was carried out using open source software. The main advantage of the ALS data is the possibility of modelling for large areas and the lack of a need for a direct visit to the measuring point. The problem is, however, the timeliness of the data. An example of this is the excision of a tree in the vicinity of one of the analyzed points, which caused gross errors in model. The second issue to deal with is difficulty in mapping the walls of buildings with ALS data. The assumption of the proposed method was the use of a raw point cloud and the maximum simplification of the modeling process. In order to cope with the problem of building walls modelling, it would be necessary to add building models from external sources or apply the algorithm of building modelling, which will be carried out in future research.

As a result of the modeling, polar plot of the elevations of the obstacles above horizon have been obtained directly from ALS data. Apart from the above-mentioned gross errors, an average error in determining the elevation of the terrain obstacles from 0.6° to 1.7° was obtained. The road equal to one elevation over the horizon is covered by the satellite in about 3 minutes, so the accuracy obtained allows for the planning of satellite visibility with an error of up to 6 minutes, which in the case of planning static GNSS sessions is a permissible quantity.

References

1. J. Jakobsen, A.B. Jensen, A.A. Nielsen. *Simulation of GNSS reflected signals and estimation of position accuracy in GNSS-challenged environment.* Journal of Geodetic Science, **5**, 1, (2015)
2. M. Bakuła, S. Oszczak, R. Pelc-Mieczkowska. *Performance of RTK Positioning in Forest Conditions.* Journal of Surveying Engineering, ASCE, **135**, 3, 125-130 pp. (2009)
3. P. Zatelli, D. D'Incà. *New modules for satellite surveying planning in GRASS.* Geomat Workb 3:1–14, (2004)
4. R. Pelc-Mieczkowska. *Primary results of using hemispherical photography for advanced GPS mission planning.* In 9th International Conference Enviromnental Enginiering, Vilnius, Lituania, (2014)
5. R. Pelc-Mieczkowska, J. Janicka, M. Bednarczyk, D. Tomaszewski. *Comparison of selected data acquisition methods for GNSS terrain obstacles modeling.* Acta Geodynamica et Geromaterialia, **12,** 3, 307-316 pp. (2015)
6. S. Ackermann, A. Angrisano, S. Del Pizzo, S. Gaglione, C. Gioia, S. Troisi. *Digital Surface Models for GNSS Mission Planning in Critical Environments.* Journal of Surveying Engineering. 10.1061/(ASCE)SU.1943-5428.0000119, SSN (print): 0733-9453, ISSN (online): 1943-5428, (2013)
7. P. D. Groves. *Shadow matching: A new GNSS positioning technique for urban canyons.* The journal of Navigation, **64**, 3, 417-430 pp. (2011)
8. P. Wężyk. *Podręcznik dla uczestników szkoleń z wykorzystania produktów LiDAR.* Główny Urząd Geodezji i Kartografii, Warszawa, (2015) (in Polish)
9. S. Xu, G. Vosselman, S. Oude Elberink. *Detection and classification of changes in buildings from airborne laser scanning data.* Remote sensing, **7**, 12, 17051-17076 pp. (2015)

Application of GIS technology in the protection of monuments on the example of Historic Monuments

Krzysztof Rząsa[1,*], *Marek* Ogryzek[1], and *Mateusz* Ciski[1]

[1]Faculty of Geodesy, Geospatial and Civil Engineering, Institute of Geography and Land Management, University of Warmia and Mazury in Olsztyn, Prawochenskiego Street 15, 10-720 Olsztyn, Poland

Abstract. According to the criteria set by The National Heritage Board of Poland, an non-movable monument of trans-regional importance, of high historical, scientific and artistic values; relevant to the Polish cultural heritage, consolidated in the social consciousness by being a source of inspiration for future generations, may be recognized as Historic Monument. As at April 20, 2018, there are 91 Historic Monuments in Poland. The objects were characterized by: location, type, age, date of recognition as a Historic Monument. Several GIS tools, such as: Kernel Density, diagram map with proportional symbols, and various forms of graduated symbols were used to analyse and visualise the spatial data. Results were presented in tabular form and on a series of created maps. The applied GIS tools allowed to elaborate and show spatial, temporal and generic differentiation of analysed objects; which allowed to draw conclusions in this area, included at the end of the article.

1 Introduction

Recognition of an object or area as a Historic Monument (pol. pomnik historii), in accordance with art. 7 of the act of 23 July 2003 on the protection of monument and the care of monuments [1], along with an entry in the register of objects of cultural heritage (pol. rejestr zabytków), an entry on the List of Heritage Treasures (pol. Lista Skarbów Dziedzictwa), Cultural Park (pol. park kulturowy) establishment and protection regulations in spatial planning documents, is one of the legal forms of protection of monuments in Poland. This form was created as a kind of ennoblement of the most valuable monuments in Poland [2], although according to B. Kwiatkowska-Kopka [3], from the formal point of view, there is no doubt that recognition as a Historic Monument is not accompanied by any support and its managers must, in fact, only rely on their own strength alone. Owners of buildings that have been ennobled and included in the list of Historic Monument do not acquire any public or legal rights. The more active owners are trying to obtain financial resources, supporting their efforts with the brand of Historic Monument. However, this does not change the fact that Historic Monuments should be supported by appropriate legal, financial and organizational solutions. According to Zalasińska [4], recognition as a Historic Monument does not entail any additional legal consequences. It does not change the legal status of the object from the point of view of the forms of the protection of monuments. Therefore, it does not give the conservation administration additional instruments of operation, and protected objects and areas are in fact subject to a preservation regime, resulting from an entry in the register of objects of cultural heritage or obtaining the status of a cultural park. The only effect is in fact an impact on social awareness. Konopka [5] indicates, that thanks to recognition as a Historic Monuments, this form has obtained a formal and permanent place in Polish law for the protection of cultural property, and Minister of Culture and National Heritage was obliged to determine the manner of carrying out the protection of Historic Monuments, which in consequence meant the right legal formula for the protection of areas. In accordance to art. 15 of the Act [1], the President of the Republic of Poland, at the request of the Minister of Culture and National Heritage, by regulation may acknowledge as Historic Monument a non-movable monument entered in the register of objects of cultural heritage, or a Cultural Park of special value for culture. Minister of Culture and National Heritage may submit the application after obtaining the opinion of the Council for the Protection of Monuments. The withdrawal of recognition of the non-movable monument as a Historic Monument takes place in the procedure provided in its recognition. Minister of Culture and National Heritage can present to The World Heritage Committee an application for entry of a Historic Monument on "World Heritage List", to protect this object on the basis of Convention Concerning the Protection of the World Cultural and Natural Heritage, adopted by UNESCO during the 17th session in Paris on November 16, 1972.

An object recognized as a Historic Monument becomes an element of cultural heritage and is subject to the principles of protection characteristic of this type of objects. Such activities can be observed all over the world [6-8].

* Corresponding author: krzysztof.rzasa@uwm.edu.pl

Of the currently existing 91 Historic Monuments in Poland, as much as 40% are objects related to religion. Culture and cultural values are commonly associated in conservation with the spiritual and religious significance of landscapes and wildlife, manifested in sacred sites or spiritual landscapes, and in rituals, customs and ceremonies [9-13].

In this article, the Authors presented the possibilities of using GIS software in analyses related to Historic Monuments in Poland as a form of the protection of monuments. The GIS contains tools for conducting analysis and spatial modelling, and allow the user to both generate new information based on existing data, and to search information in multiple layers of data. GIS also allows the visualisation and analysis of the data cross-referenced by time (which shows time trends), and for the data to be used for the purposes of protection of monuments.

2 Materials and methods

To conduct analyses using GIS software, in the first step, a detailed analysis of legal grounds of functioning of Historic Monuments in Poland and detailed inventory and classification was made. According to the criteria set by National Centre for Research and Documentation of Monuments (currently The National Heritage Board of Poland, pol. Narodowy Instytut Dziedzictwa), an non-movable monument of trans-regional importance, of high historical, scientific and artistic values; relevant to the Polish cultural heritage, consolidated in the social consciousness by being a source of inspiration for future generations, may be covered by this form of additional protection. It is advisable for these monuments to: maintain the original spatial composition or have undergone slight transformations, be uniformly stylish or have clear and harmonized layers, be properly exposed in urban space or landscape, retained their original relationship with the environment, be the works of outstanding artists, e.g. architects, planners, landscape architects, gardeners; be well preserved or in a condition allowing their revalorization, and be the subject of conservation care. From a formal point of view, a regulation of the President of the Republic of Poland is required (the procedure indicated in the Introduction). Such criteria as of 20/04/2018 meets 91 objects. These are various objects that, in order to improve further analysis, were divided into 9 generic-functional groups: sacral and sepulchral units (36 objects), urban systems (16), residential units (11), monuments of technology (10), works of architecture and construction (6), cultural landscapes (4), archaeological monuments (3), defensive structures (2) and battlefields (3). For all objects: the dates of creation, location in a given voivodeship, detailed location and date of recognition as a Historic Monument were collected. On such a prepared database, analysis using GIS applications were carried out. The planned analyses required precise point data, presenting the location Historic Monuments in Poland. Therefore, the available online GIS services were analysed, in terms of the ability to automatically download the necessary data. The only service that met this requirement was the service DanePubliczne.gov.pl. This service is designed to collect in one place data of special importance for the development of innovation in the country and the development of the information society. Data downloaded in CSV format (with columns containing geographical coordinates) was converted to point data using the "Make XY Event Layer" tool, from the ArcMap component of the ArcGIS software. The acquired data set has been updated and supplemented with 10 objects, recognized as Historic Monuments on 20/04/2018.

If we assume that the location of Historic Monuments is associated with territorial division units [14] (in the case of a this analysis – voivodeships), presentation of data in the form of a cartogram (Fig. 1) will show the basic way of data distribution. According to the classification of the map, a darker grey shade means more Historic Monuments in a given voivodeship.

To present the year of recognition of the object as a Historic Monuments, graduated symbols (Fig. 2) was used – data has been divided into four time intervals, depending on the date of recognition (up to 1994, 1995-2007, 2008-2012, 2013-2018), which were marked with gradually darker points (which means more recent Historic Monuments). However, more recent Historic Monuments do not necessarily mean later created objects – Fig. 3 in a similar way as Fig. 2 (with additional geometric signatures) shows the age of objects recognized as a Historic Monuments.

In order to visualize 9 generic-functional groups into which Historic Monuments were divided, a diagram map (cartodiagram) was created (Fig. 4). Circular diagrams with the numbers of Historic Monuments from individual generic-functional groups were drawn on the borders of all voivodeships. In addition, the size of the diagram shows the total number of Historic Monuments in a given voivodeship.

Using the Kernel Density function (Fig. 5), the density of the variable distribution was determined based on known locations of Historic Monuments in Poland 15]. Due to the domination of religious objects indicated in the Introduction, to study the spatial distribution of the types of Historic Monuments, data on generic-functional groups from four main groups was selected (sacral and sepulchral units – 36 objects, urban systems – 16, residential units – 11, monuments of technology – 10), and then the Kernel Density function was used again to create four another maps (Fig. 6. Sacral and sepulchral units, Fig. 7. Urban systems, Fig. 8. Residential units and Fig. 9. Monuments of technology).

3 Results and discussion

Recognition the first map presents distribution of Historic Monuments in Polish voivodeships (Fig. 1). Cartogram (with added points with exact location of monuments) presents south-western Poland as rich in Historic Monuments. On average, one voivodeship has 6 historical monuments, while only the Greater Poland and Lower Silesian voivodeships have more than 9 (10 and 12 respectively).

Analysis of the year of recognition of the object as a Historical Monument is shown in Fig. 2. The largest and darkest symbols mean objects that are later recognized as a Historic Monument (monuments recognized in the years 2013-2018 constitute approx. 40%).

In order to compare the age of the object to the year it is recognised as a Historic Monument, in a similar manner as in Fig. 2, Fig. 3 shows the four ranges of centuries of the objects' creation (from BCE, I-XII century, XIII-XVIII, and XIX-present).

The cartodiagram (Fig. 4) depicting the division of Historic Monuments into generic-functional groups shows very well the domination of religious objects in almost all of Poland.

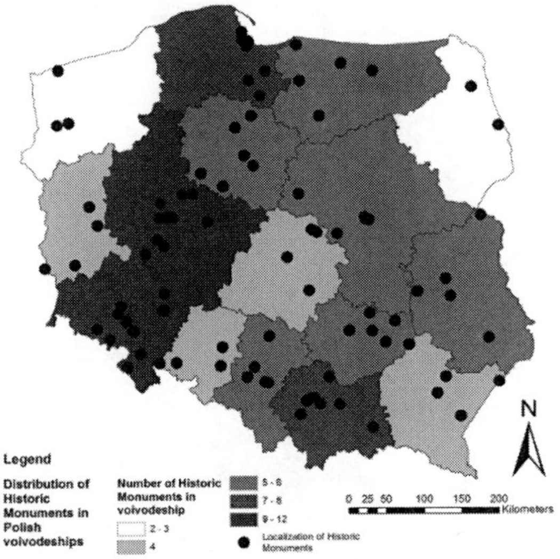

Fig. 1. Distribution of Historic Monuments in polish voivodeships.

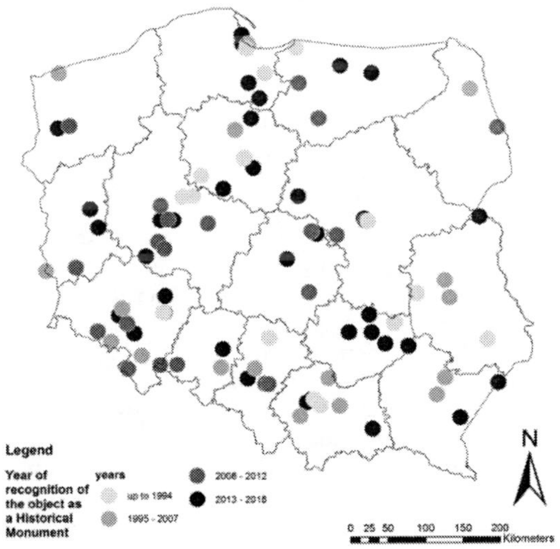

Fig. 2. Year of recognition of the object as a Historical Monument.

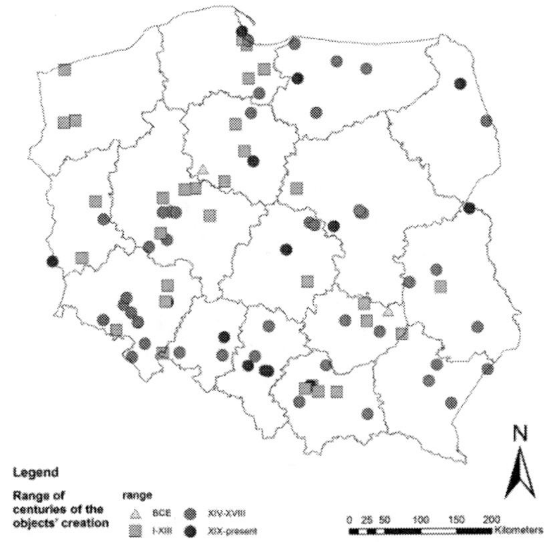

Fig. 3. Range of centuries of the objects' creation.

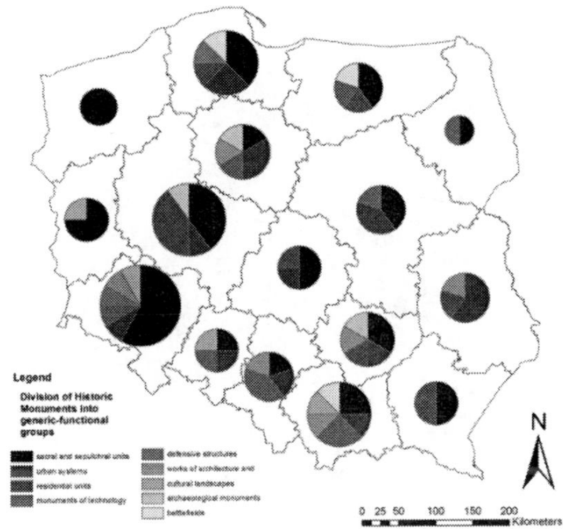

Fig. 4. Division of Historic Monuments into generic-functional groups.

Analysis of the Kernel Density function (Fig. 5) more accurately than the cartogram illustrates the density of the location of Historic Monuments in Poland. By removing the limitations in the form of administrative boundaries, the natural, spatial distribution of historical monuments was examined [14,15]. Fig 5 presents several clusters of historical monuments: the vicinity of the Tri-City, Poznań and Wrocław.

In order to examine the influence of various generic-functional groups on the density of the location of Historic Monuments, following maps show the density of the distribution of groups: Fig. 6. Sacral and sepulchral units, Fig. 7. Urban systems, Fig. 8. Residential units and Fig. 9. Monuments of technology.

https://doi.org/10.1051/e3sconf/20186300015

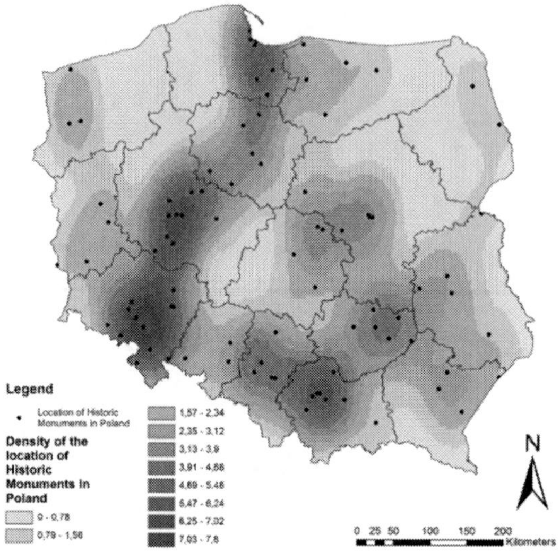

Fig. 5. Density of the location of Historic Monuments in Poland.

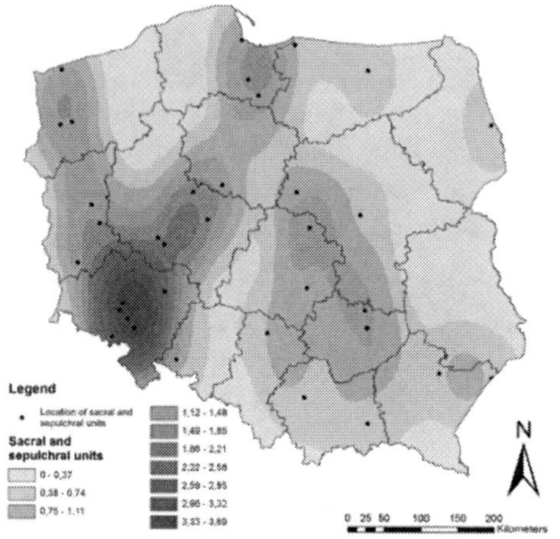

Fig. 6. Sacral and sepulchral units.

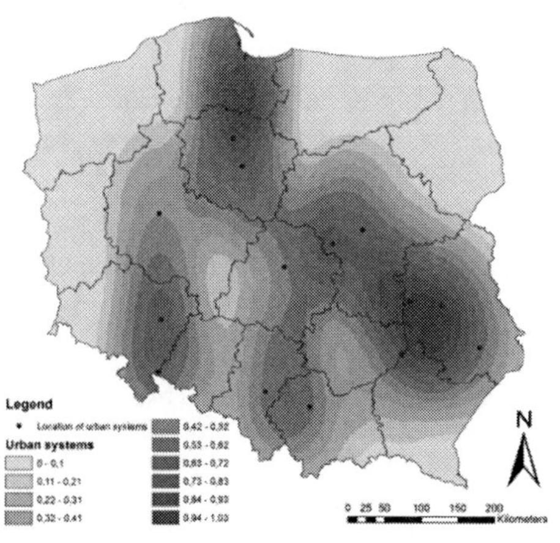

Fig. 7. Urban systems.

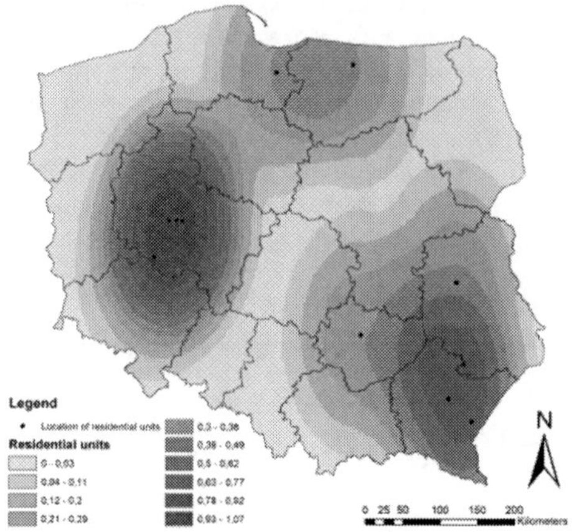

Fig. 8. Residential units.

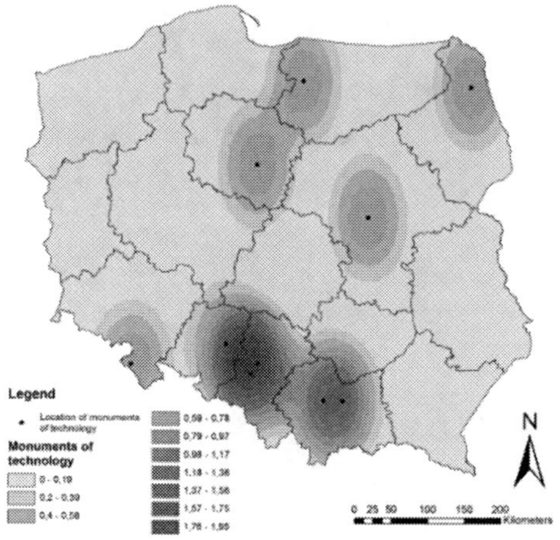

Fig. 9. Monuments of technology.

4 Conclusions

Analyses carried out allowed to formulate the following conclusions:

- GIS technologies gives a wide range of spatial analysis capabilities. Presented methods of spatial analysis are a very good tool in the analysis of monuments
- Websites and online databases are an excellent spatial data resource. Necessary data on Historic Monuments (resulting from the applied methodology) are available for free download, automatically.
- Occurrence of Historic Monuments does not seem to correlate with specific voivodeships – a few large clusters can be noticed, but there is no continuity and parallel distribution. The locations corresponds with historically Polish areas (with the exception of the Lower Silesian voivodeship, but in this case more than half of the Historic Monuments are sacral and sepulchral objects).
- The conclusion from point 3 is additionally visible after applying the division into the age range – the oldest objects (BCE and I-XIII) they dominate in the south-western part of the country, while newer ones (interval

XIV-XVIII definitely, XIX-present range – less) cover the entire surface of the country without any visible trends.

- Partitions of Poland and the history of the country is indirectly visible on the maps – there is an accumulation of Historic Monuments on historically Polish lands (Greater Poland, Pomeranian, Mazovia, Subcarpathia), but also on the Recovered Territories (Lower Silesia)
- Representing 40% of all Historic Monuments, sacral and sepulchral objects on Fig. 6 fit well into the overall density shown in Fig. 5. Residential units (Fig. 8) are responsible for aggregation around the city of Poznań. Monuments of technology (Fig. 9) occur almost only in the south of the country, and urban systems (Fig. 8) seem to be responsible for the rest of the country (map is almost the exact opposite of sacral and sepulchral units).
- Urban systems (Fig. 8) are closely related primarily to the historical cities of Poland.

References

1. The Act on the Protection of Monuments and the Care of Historical Monuments. Parliament. Journal of Laws from 2014, No 1446 as amended (2003)
2. H. Kisilowska, *Prawne aspekty ochrony nieruchomości zabytkowych*, Mazowsze, Studia Regionalne, Mazowieckie Biuro Planowania Regionalnego w Warszawie, No **7**, pp. 59-65 (2011) (in Polish) DOI:10.21858/msr
3. B. Kwiatkowska-Kopka, *Pomnik historii – i inne formy nobilitacji zabytku jako priorytety na rzecz powszechnego dostępu do dziedzictwa i jego promocji*, Wiadomości Konserwatorskie 2015 No **44**, pp. 98—109 (2015) (in Polish) DOI:10.17425/WK44MONUMENT
4. K. Zalasińska, *Pole bitwy pod Grunwaldem pomnikiem historii*, Spotkania z zabytkami, **3-4** marzec-kwiecień 2011, pp. 47–48 (2011) (in Polish)
5. G. Mikusiński , H.P. Possingham & M. Blicharska, *Biodiversity priority areas and religions—a global analysis of spatial overlap*. Oryx, The International Journal of Conservation, published for Fauna & Flora International. No **48**, pp. 17–22 (2014)
6. P. M. Protopsaltis. *Book Review: Cultural Heritage in International Investment Law and Arbitration, by Valentina Vadi*. (Cambridge: Cambridge University Press. 2014). 35 *Journal of International Arbitration*, Kluwer Law International BV, The Netherlands. Issue **2**, pp. 267–271 (2018)
7. E. Lagrange, S. Oeter, R. Uerpmann-Wittzack, eds. *Cultural Heritage and International Law: Objects, Means and Ends of International Protection*. Springer International Publishing, Basel. (2018) DOI:10.1007/978-3-319-78789-3
8. X. Jiao, J. Wang. *Context Protection and Inheritance in the Process of Urbanization*. 2nd International Conference on Education, Management and Applied Social Science (EMASS), pp. 170-174 (2018) DOI:10.12783/dtssehs/emass2018/20409
9. K. Metcalfe , R. Ffrench-Constant & I. Gordon. *Sacred sites as hotspots for biodiversity: the Three Sisters Cave complex in coastal Kenya*. Oryx, The International Journal of Conservation, published for Fauna & Flora International. No **44**, pp. 118–123 (2010)
10. N. Gupta, A. Kanagavel, P. Dandekar, N. Dahanukar, K. Sivakumar, V.B. Mathur & R. Raghavan. *God's fishes: religion, culture and freshwater fish conservation in India*. Oryx, The International Journal of Conservation, published for Fauna & Flora International. No **50**, pp. 244–249 (2016)
11. E.P. Riley. *The importance of human–macaque folklore for conservation in Lore Lindu National Park, Sulawesi, Indonesia*. Oryx, The International Journal of Conservation, published for Fauna & Flora International. No **44**, pp. 235–240 (2010)
12. J.E. McKay, F.M. Mangunjaya, Y. Dinata , S.R. Harrop & F. Khalid, *Practise what you preach: a faith-based approach to conservation in Indonesia*. Oryx, The International Journal of Conservation, published for Fauna & Flora International. No **48**, pp. 23–29 (2014)
13. G. Mikusiński , H.P. Possingham & M. Blicharska. *Biodiversity priority areas and religions—a global analysis of spatial overlap*. Oryx, The International Journal of Conservation, published for Fauna & Flora International. No **48**, pp. 17–22 (2014)
14. B. Medyńska-Gulij, *Kartografia i geowizualizacja*. Warszawa: Wydawnictwo Naukowe PWN (2011) (in Polish)
15. P. A. Longley, M. F. Goodchild, D. J. Maguire, and D. W. Rhind, *Geographic Information Systems and Science*, 2nd Revise. John Wiley & Sons Ltd., New York (2005)

Flood Hazard Zone Modeling for Cadastral Plots in Ojców National Park in Poland

Izabela Piech*, *Stanisław* Bacior, *Monika* Mika

University of Agriculture in Krakow, Faculty of Environmental Engineering and Land Surveying, Poland

Abstract. The paper presents the stages of development of DTM for fragment of ONP (Ojcow National Park). Model was created in order to illustrate potential flood zones of river Pradnik. In subsequent stages of the development are shown: Basic concepts of subjects taken in the publication, characteristics of research object, methodology of DSM implementation, elements of flood zones and assessment of flooded plots devastation impact on their cadastral and landscape values. The advantage of the method is a low cost and short time of DTM creation based on aerial photographs. Created model has also a relatively high accuracy of elaboration. It is noteworthy that for the area being elaborated independent survey were performed using classical and modern technology methods. The work results will provide a base for further analysis in order to determine the flood zones for mountain rivers. Described problem will be the subject of subsequent papers by the authors.

1 Introduction

The aim of the publication is to describe the implementation methodology of DTM and DSM in ONP in Poland, with particular emphasis on the determination of the flood zones of the river Prądnik. The study area includes the Prądnik Valley. The coastline of the river that is the subject of research is variable along its length. It does not coincide with the evidence borders of the river. The specific character of field surveying work in ONP requires from the surveyor not only knowledge of surveying law, but also knowledge of the constraints in national parks. Methodology of classical field work consists in the use of different measurement techniques and the integration of acquired data. The terrain configuration makes it necessary to use different techniques and surveying equipment (e.g., tacheometry, GNSS). Presented in the paper method of relief elaboration did not require field survey, except an interview in order to determine the control photo points. The model was made on the basis of the pictures from photogrammetric flight. This method appears to be much more beneficial in terms of economic (cost and time) to develop. Accuracy of the method is more than sufficient to estimate the range of flood zones for flood protection of the selected area [1,2]. In the study the following research tools were used: digital photogrammetric station DELTA digital photogrammetric station DEPHOS, SURFER program, Microsoft Excel, and data from the Geoportal.

2 Terminology from the scope of the research

In the international terminology the Digital Terrain Model is defined as: "Numeric representation of the terrain surface, formed by a set of properly selected points on the surface and interpolation algorithms allowing to render it in a specific area" [3].

According to [4] of the fundamental problems associated with the digital terrain model are:

• the sampling problem - problem of proper selection of the characteristic points of the surface in order to get the best results while minimizing the amount of data,
• representation problem - the problem of surface reproduction on the basis of existing data.

In turn, from a technical point of creating DTM there are two basic types of models [4-7]:

• GRID model - in the form of a regular grid of squares supplemented by the characteristic points and skeletal lines,
• TIN model - in the form of an irregular triangle mesh.

Data to create a digital terrain model derived primarily from three sources [8, 9, 11, 12]:

• direct measurements,
• photogrammetric measurements,
• digitization of existing maps.

Occasionally also radar or laser altimetry are used. These techniques are mainly used for studies of geological models, drilling or seismic measurements.

The term flood zone means the extent of flood waters, determined on the basis of long-term historical data or theoretical considerations, that determine the probability of flood events or local flooding in the area [13]. There are direct and potential areas of risk.

* Corresponding author: rmpiech@cyf-kr.edu.pl

The first is the area immediately adjacent to rivers or streams and areas in the immediate vicinity of levees, where there is a threat that water on a higher level than their crown can pour over them. In turn, the potential risk areas include the areas exposed to the destructive activities of water in the case of loss or substantial damage to the embankment [14].

In the last decade in Poland, there has been an increase of rainfalls in the months from May to July. Most of the rainfalls are violent storms bearing damages of local range. Therefore, the municipalities in collaboration with the staffs of the flood hazard and public emergency services have made efforts to obtain in a short time graphic materials necessary to elaboration potential flood zones. Flood hazard maps in Poland are above all used for analyses of the risks and they facilitate local government decision-making on mitigating the effects of floods, if there are any [15]. The first phase of flood hazard mapping concerns acquisition of the historical data related to maximum river or watercourse floods. The second stage is to determine the water level elevation to receive as a result the profile for a given flow. Flood zone boundary is determined by the intersection of the water surface profile for a given flow rate and the Digital Terrain Model obtained e.g. by photogrammetric methods [16, 17]. The resulting zones are also divided into the potential and direct danger zones. Final flood hazard maps are made on the basis of raster maps, surveying profiles and knowledge of hydrology [18,19,20].

The smallest component of the map, showing the unit of land is a cadastral plot. The concept of a plot of land in Poland is defined by [21]. According to the Regulation on the registration of land and buildings a cadastral plot is a continuous area of land situated within the cadastral district, legally homogeneous, separated from the environment by means of boundary lines. Continuity (area consistency) of a plot means that it cannot be composed of two or more separate surface objects. Location within the cadastral districts guarantees that a graphic view of the plot is possible to be presented on a single sheet of cadastral map. In the case of extended objects (railroads, streams) parcels are determined in such way that their boundaries coincide with the boundaries of cadastral district. Legal uniformity of land is considered only within the visualized property rights in land and buildings (e.g. ownership, perpetual usufruct). The plot is defined in terms of geometry by its separation from the environment by boundary lines connecting borders breakpoints. Physically, in the boundary break down points boundary markers are placed and limit lines can be represented by objects like fences, walls of buildings, etc. By determining the boundaries of the cadastral plots, in the case of mutual intersection of railway lines, public roads, internal sea waters and inland flowing waters, the following rules are applied:

• internal waters and inland waters flowing, with the exception of water flowing through pipelines or covered channels divide railways and public roads into separate cadastral plots photogrammetric measurements,
• railway lines divide public roads into separate cadastral plots,
• public roads of higher category divide roads of lower category into separate cadastral plots.

The concept of the cadastral parcel is inseparable connected with the concept of the land use consistent mostly with the local land use plan or community development study [22].

According to the previously cited Regulation of the Minister of Regional Development and Construction dated. March 29, 2001 on the land and buildings cadaster – usage of land can be described as: agricultural land, forest land and wooded and shrub land and urbanized, land under water, orchards of the area less than 0.1000 ha and other land of the area. less than 0.0100 ha which is not shown in cadaster. The former can be divided into: "Arable land, orchards, durable meadows, permanent pastures, agricultural land built-up, land for ponds and ditches. Built-up areas and urbanized area are divided into: residential areas, industrial areas, other built-up areas, urban area, recreational areas, land mines and transportation land".

The quoted provision also defines arable land, which include land mechanically cultivated for production purposes, land used for the cultivation of wicker, hops, ornamental trees nurseries and fallows. Orchards - land of the area greater than 0.1000 hectares planted with fruit trees and bushes, fruit trees nurseries and vineyards. Permanent grassland is generally covered with long-term land vegetation such as grasses, legume plants and herbs, regularly mowed. Permanent pastures, as meadows, are covered with long-term plants, with the difference that they are not mowed but grazed. To the built-up agricultural land we include land occupied by residential buildings and other buildings used for agricultural production, with certain exceptions. Land under ponds are areas for reservoirs of water that are used for breeding and keeping fish (exceptions: lakes and retention reservoirs). The land under the ditches we include land occupied by drainage ditches.

3 Characteristics of the investigated object

The study area covers a part of the Prądnik river valley in the Ojców National Park and the agricultural area lying south of it. This area has an attractive landscape, and is lying close to the routes of historical Krakow in Poland. Valley is composed mainly of limestone from the period of the Upper Jurassic and Cretaceous. They were formed 150 million years ago, and their thickness reaches up to 200 m. In some places they are covered with a layer of loess. In the limestones karst features developed very well, the effect of which is the creation of numerous caves (about 300), gorges and ravines with exsurgents, rock gates, spiers and island mountains of fantastic shapes. Rock formations protruding above the surface are composed of rocky limestone with high hardness and compactness, located between the softer limestone layers. The most common fossils are sponges and brachiopods, very seldom ammonites.

The valley is the result of the erosive action of water flowing on its bottom. On the slopes of the valley terraces formed at different times of the formation of the valley can be observed. In Pieskowa Skała they are 12 and 40 m above the valley floor, in Ojców 30-40 m and 80 m. On these terraces rise Ojców Castle and Pieskowa Skała Castle.

Prądnik Valley is widely recognized as the most beautiful valley of the entire Jura. Its biggest attractions are:

- Pieskowa Skała Castle and Ojców Castle,
- available to tourists Dark Cave with the Neanderthal campsite,
- rocks, among others: Hercules' Club, Wernyhora Rock, Deotyma Needle, Maiden Rocks, Glove, Łaskawiec, Cracow Gate,
- Museum of the Ojców National Park,
- Chapel "On The Water".

In the analyzed area of research can also be seen predominance of the summer precipitations over winter ones (Fig. 1). The months June, July and August (indicated in Fig. 2) are as much as 40% of annual precipitation. Recent years show a minimal decrease in the amount of precipitation. These statistics do not, however, take into consideration floodings occurring locally in the area as a result of extreme weather.

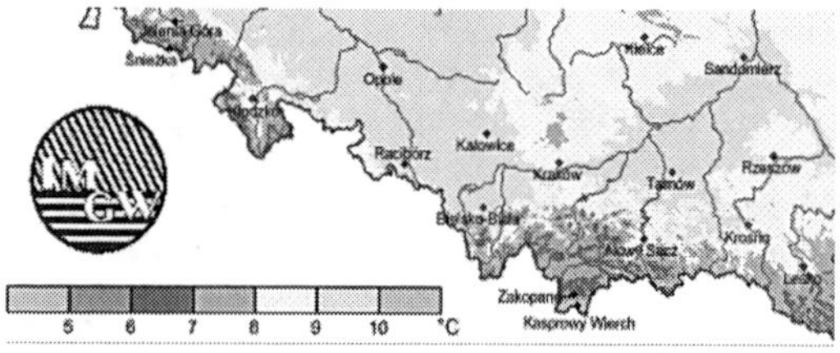

Fig. 1. Map of the temperature distribution of the studied region of Poland.

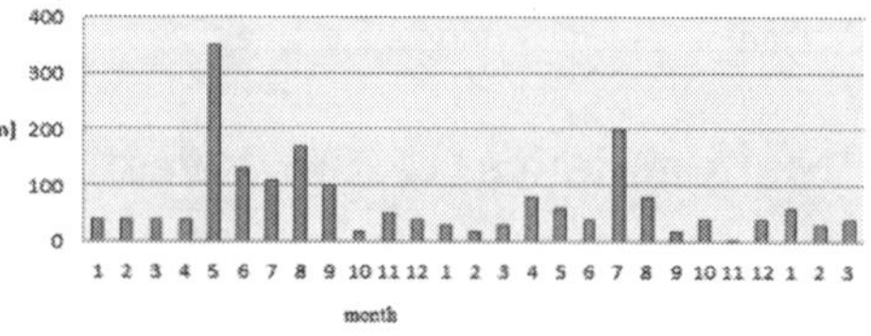

Fig. 2. Total rainfall for the investigated area in the years 2010, 2011 and 2012, drawn up on the basis of Polish climatic maps from the Institute of Meteorology and Water Management.

4 Detailed methodology for the development of DTM and DSM

4.1 The application of digital station "Delta"

DTM is based on aerial stereograms in a scale 1:13000, made by RC20_13166 analog camera. Aerial photos in digital form were obtained from the Marshal's Office in Krakow. A digital photogrammetric station "Delta" was used for the elaboration. Below the various stages of the work with the use of specified materials are described.

4.2 Creation of a project and data entry

The development started with the creation of a new project in the Camera lap. Next data from metrics measuring camera were introduced (name of the camera, the focal length, the base of shooting, shifting the main coordinates of background stamps and the distortion). On the Ground lap control photo points coordinates, obtained from geodetic survey were entered. Due to the insufficient number of control points, for elaboration were used additional points derived from aerial triangulation or chamber densification of geodetic network. Obtaining of coordinates of the points was possible by the use of Dephos 4.10 software (Fig.3). The program initially converted a pair of images, then in the application Dephos Manager 4 batch file provided with aerial triangulation was placed, then using the Mapper Stereo lap all the points from the chamber densification of geodetic network were visualized.

Fig. 3. Models Module Interface.

Then, in the application DEPEdit a pyramid of images was created. After entering the required data in the Models module - orientations of the photographs (internal, mutual, absolute) and stereograms vectorization (Mapping) were performed (Fig.4).

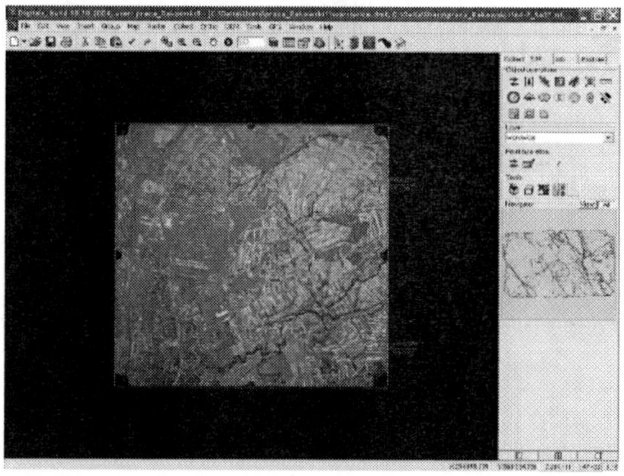

Fig. 4. Vectorization of the stereogram using digital station "Delta".

4.3 Performance of images orientation

Internal orientation, aimed to transform coordinates from the pixel system to the fiducial coordinates system was performed basing on previous data entry in the Camera lap. This orientation is done on a two-dimensional model, it consists in the location of the apparent measurement mark in the middle of the picture of fiducial mark. To proceed with the orientation it is

https://doi.org/10.1051/e3sconf/20186300016

necessary to from the Models Module the Models Model lap, and then inner orientation. The first step is to set the mark at the appropriate point in the left image, then hold down the Ctrl key, and repeat this action in right image. Pressing the right mouse button confirms the position of "apparent measurement mark". Photogrammetric Station itself imposes order to carry out this operation. After setting all the points, we get a report that indicates the accuracy, if it is too low, it is possible to improve a particular point until you reach the minimum permissible errors. Next, a mutual orientation of the images was performed. This activity was aimed at orientation pictures with each other, what is obtained by removing transverse parallax and the transformation of the fiducial coordinates to the model coordinates.

From a technical point of view, acceding to complete this orientation, it is necessary to select the Model lap, and then relative orientation. The standard measurement is performed on six points, called Gruber points, the best area of the points distribution is indicated by the photogrammetric station. And the operator decides on the selection of specific points, it must be remembered that these points were in proposed by the software area and that they do not coincide with the photo points. Orientation results are written to the report.

In the next stage of the work an absolute orientation of the pair of images was performed. This activity was necessary to transform coordinates from a model system to the surveying system, in which further measurements were made. This orientation is done using minimum three control points which have be input in the program Ground Models lap.

In order to improve the accuracy of orientation relative to the geodetic system, the absolute orientation of this study used four control points. Implementation of absolute orientation was performed on a three-dimensional model. First the Model lap was chosen, then the absolute orientation. Aerial photo (aerial) measurement method and the stereo model method of orientation were selected. In the beginning it is important to set the glasses in stereo mode (for this purpose it is necessary to select from control the menu, and then OpenGL Glasses). Using the "apparent surveying mark" the situational and altitude location is determined for specified by the program GSP fotopoints.

An important step for the preparation of DTM was data vectorization. It was made using the Mapping lap. First, the relevant thematic layers were created together with a choice of relevant attributes, such as color, thickness and style of the line.

The following thematic layers were created: the scope, forests, roads, water, buildings, meadows and pastures, slopes and contours. Layers of Digital Elevation Model and Triangulated Irregular Networks are created automatically when the relevant program functions are applied. The stereogram vectorization was carried out using a three-dimensional model. To apply it is necessary to choose Stereo Glasses (Open GL) in the program options in the Hardware tab. The final result of the work was vectorization of an area of approximately 1665 hectares. The next stage of the study was to create DTM. Subsequently the triangle mesh – TIN was generated. In creation of a grid of triangles all the layers that have been previously established are involved. In order to take them into account, in the window of management the layers the status must be selected to edit, then go to the Edit lap and select all objects vectorized. Meshing TIN can be created on the basis of DEM options - Create TIN. Irregular triangular mesh is generated on an automatically created new layer named: Triangulated Irregular Networks.

The last phase of work using digital station Delta was the creation and smoothing of the contours. Generating of contour lines is executed after using Raster DTM option - Build Contour lines and passing the appropriate contour lines interval contour lines. The project used cutting interval of 2.5 m. Created contours shpulb smoothed with the Smooth command, and then removed from the built-up areas, forests, slopes, roads and waters.

5 Creation of a model of the terrain and flood zones models using the surfer software

Surfer is a full-function 3D visualization, contouring and surface modelling package by Golden Software, that runs under Microsoft Windows. It is a grid-based graphic program interpolating irregularly spaced XYZ data into a regularly spaced grid [23]. In this elaboration all vectorized on photogrammetric station "Delta" situational details make the model, namely forests, meadows and pastures, built-up areas, water, roads and slopes. Digital Station allows saving the project to a text file in which the vectorized objects are being broken down into sections with appropriate geodetic coordinates. The structure of the text file generated by the "Delta" needs improvement, because in each layer there are listed objects and point coordinates are in them. Surfer in order to begin creation of the model, needs a continuous record of points separated by a space or a tab. The removal of unnecessary lines was performed using Microsoft Excel. Terrain model was developed on the basis of 27187 points. Model of the river was built on the basis of points obtained from the digitalization, plus the corresponding value of flood zone - 2 and 5 m. In the window it is necessary to choose the right method to create a grid, for this project the radial method was used. In the elaboration three models of grids were created: for the land surface, for the rivers with the water level higher by 2 m and for the river in the state higher by 5 m. All grids were smoothed using a Gaussian filter. Creation of the 3D model is done using the option New 3D Surface. After its activation, it is necessary to enter the path to the smoothed and improved grid squares. The Surfer enables application and penetration of several models, but note that the new model must be located in an existing map (Fig. 5). After the models are created it is possible to improve the graphic and descriptive elements. Fig. 5 shows the DTM with the flood zones, created for the study area.

https://doi.org/10.1051/e3sconf/20186300016

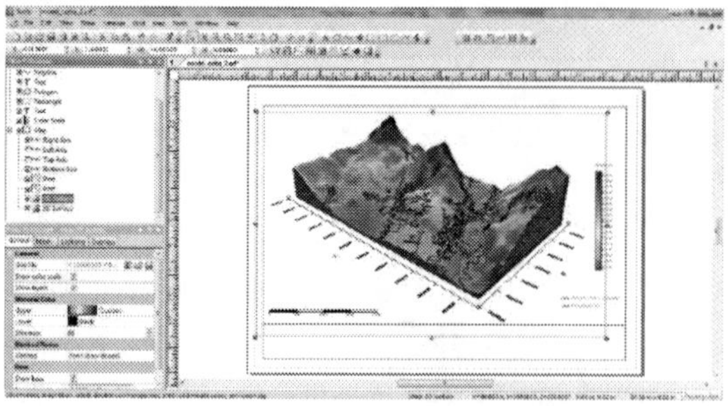

Fig. 5. Creation of the 3D visualization of the terrain model.

6 Influence of destruction caused by the high water on the cadastral and the landscape value of the flooded areas

The aim of the study was achieved through the creation of DTM and determining areas of potential flood risk in the Valley Prądnik. On the basis of generated model the areas most vulnerable to the destructive effects of an increase in water level of the river about 2 m and 5 m were estimated.

Flood risk affects more than 80% of the cadastral plots located on the border or in the immediate vicinity of the river Prądnik. Detailed results will be the subject of a separate publication.

The effects of flooding can have an irreversible, degradation impact on the landscape and cadastral values of flooded areas. This applies to both arable land and many existing historic architectural and engineering objects or the unique forms of the natural landscape in the Ojców National Park.

Particularly at risk are the areas in the vicinity of the spring of the river Prądnik in the cadastral commune of Suloszowa and the Zielonka commune. On the other hand, in the Skała commune, where the river has a very irregular coastline surrounded by lush greenery and is located between the high limestone cliffs, the most vulnerable to damage are surroundings of the Chapel "On The Water".

The subject of the research was a selected section of the Prądnik river. A comparison was made of cadastral maps, maps from the Austrian cadaster and land reclamation maps. The changes the course of the riverbed for over 100 years were examined. The study was based on overlaying rasters of cadastral and reclamation maps, and then redrawing the course of the riverbed basing on the borders of cadastral plots. Subsequently those data were compared with the course of the riverbed according to aerial photographs posted on the site geoportal2.

The examined section of the Prądnik river of approx. 2.5 km is located in the Zielonki and flows through the villages Pękowice and Zielonki. Over the entire length the section, the riverbed strongly meanders. On the definitely greater part of the study area around the river there are mostly farmland, and the course of river is natural.

The study showed that the riverbed has changed quite significantly. Definitely the largest displacements in comparison to the current course of the river are shown on Austrian cadastral maps. Differences in some places reach up to 35 m. This is due to the fact that these maps were created at the earliest and show the course of over 100 years before. Reclamation maps showed smaller differences. In contrast, the riverbed on new cadastral maps almost perfectly coincides with the riverbed shown in aerial photographs on the geoportal2 site which indicates on a relatively low activity of the river.

Prądnik river is considered one of the few rivers which run is natural almost on its entire length (except for the section flowing through the territory of the city of Krakow). Although looking at the fragment of the river shown in Fig.7 it can be concluded that this part of the river was changed as a result of human activities. In the aerial photo it can be seen, that in the place where previously there was the riverbed, there is no trace, that it ran at the original site, and at the moment there is arable land there. Therefore, it can be concluded with a high probability, that the course of the riverbed has been artificially regulated. While at the remaining part of the test section the flow changes of Prądnik are not so significant and are associated with natural processes.

One of the consequences of the fact, that the river is not in any way regulated, is the increasing siltation of the bottom, and thus reduction of its depth. The result is a quite often occurrence of local floodings, caused by sudden and heavy precipitations. They occur most often just in the Zielonki commune.

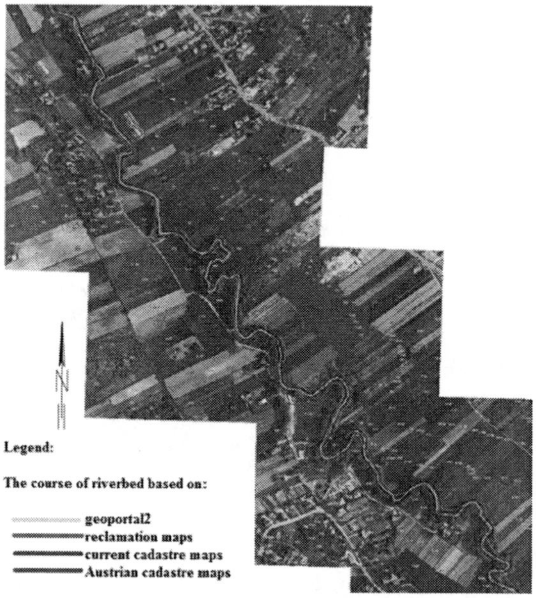

Fig. 7. The course of the Prądnik river on the test section on orthophotomap from geoportal2 [24].

7 Conclusions

The advantage of using photogrammetric methods is the possibility of work in chamber conditions, which allows independence from weather conditions. Making aerial photographs although preceded by very long preparation, provides a large amount of materials giving coverage over areas such as several communes. The accuracy of the measurements to create a Digital Terrain Model is very high. Progress has been made in the construction of the measuring cameras allows to perform photogrammetric digital images that are free from errors caused by photos scanning.

Due to the location of many of the built-up areas in river valleys, the definition of flood zones based on digital terrain model with elements of its coverage, becomes the basis for municipalities and counties to counteract the effects of floods.

Basing on the described methodology for the creation of the DTM simulation of the flood zones can be made for each commune, as well as full monitoring of vulnerable areas.

References

1. M. Kemblowski, et al. *Risk modelling with Bayesian Networks – case study: construction of tunel under the Dead Vistula River in Gdansk*, Procedia Engineering, **196**, pp. 585-591 (2017)
2. Regulation of the Minister of Internal Affairs and Administration dated November 3, 2011r. regarding the databases on aerial and satellite imagery and orthophotos and digital terrain model (Journal od Laws No. 193, item 1287) (2011)
3. R. Weibel, M. Heller, Digital terrain modelling. In Maguire, D. J., et al. (eds.), Geographical Information Systems: Principles and Applications. London: Longman (1991)
4. I. V. Florinsky *Errors of signal processing in digital terrain modelling* International Journal Geographical Information Science Volum 16, pp.475-501 (2010)
5. J. Kocyła, *Visualization, methodology and proposals for the use of the numerical terrain model in geology on the example of the spatial model of the Warta River section,* Warszawa: Surveying Review 45, pp. 211–214 (1997) (in Polish)
6. K. Kraus, J. Otepka,. *NMT Modelling and Visualization* – The SCOP Approach. 'Photogrammetric Week 05', Wichmann Verlag, Heidelberg, pp. 241-252 (2005)
7. K. Pyrchla, A. Inglot, J. Pyrchla, *Fuzzy Sets in the GIS Environment in the Location of Objects on the Surface of Water Bodies*, DOI: 10.1109/BGC.Geomatics.2017.44, Publisher:IEEE, BGC Geomatics (2017)
8. O. Dorozhynskyy, T. Wrona. *Basics of photogrammetry,* Publisher of the Lviv Polytechnic, Kraków – Lwów (2003)
9. N. El-Sheimy et al. *Digital Terrain Modeling: Acquisition, Manipulation And Applications*, Artech House (2005)
10. Braun, H. McQueen, M. Sambridge. *Geophysical parametrization and interpolation of irregular data using natural neighbours*, Geophys. J. Int. **122**, pp. 837–857 (1995)
11. K. Bobkowska, et al. *Implementation of Spatial Information for Monitoring and Analysis of the Area Around the Port Using Laser Scanning Techniques*, PMR, **24**, pp 10-15 (2017)
12. A. Janowski, et al. *Airborne and mobile laser scanning in measurements of sea cliffs on the southern Baltic*, GeoConference-SGEM (2015)
13. Directive 2007/60/EC of the European Parliament and of the Council of 23 October 2007 on the assessment and management of flood risks (Official Journal of the European Union, L288), pp. 27 (2007)
14. Z. Zwoliński, *On the homology of geoinformatic terminology. [In:] GIS - water in the environment,* Poznań: Bogucki Scientific Publisher, pp. 21-30. ISBN 978-83-6132091-3 (2010)
15. Water Law dated July 20, 2017 (Journal od Laws 2017, item 1566), (2017)
16. M. Szydlowski, W. Szpakowski, P. Zima, *Numerical simulation of catastrophic flood: the case study of hypothetical failure of the Bielkowo hydro-power plant reservoir*, Acta Geophysica, **61**(5), pp. 1229-1245 (2013)
17. L. Hai-Min et al., *Flood risk assessment in metro systems of mega-cities using a GIS-based modelling approach*, SC OF THE TOT ENVIR, **626**, pp. 1012-1025 (2018)
18. E. Drożdżal, *Flood hazard zones*, Gospodarka wodna, **11**, pp. 425 – 427 (2006)
19. V. B. Mitkova et al. *Reconstruction and post-event analysis of a flash flood in a small ungauged basin: a case study in Slovak territory,* Natural Hazards, **92**, pp. 741-760 (2018)
20. W. Szpakowski, M. Szydlowski, *Evaluating the Catastrophic Rainfall of 14 July 2016 in the Catchment Basin of the Urbanized Strzyza Stream in Gdansk, Poland,* Pol Jour of Envir St, **27**, pp. 861-869 (2018)
21. Regulation of the Minister of Regional Development and Construction dated. March 29, 2001 on the registration of land and buildings (Journal od Laws No. 38, item 454), (2001)
22. P. van Oosterom , E. Fendel *Proceedings of* 4th International Workshop on 3D Cadastres, Dubaj, ISBN: 978-87-92853-21-8 (2014)
23. J. Redl, V. Valikova, J. Antl, *Modelling of Terrain Surface* Acta Technologica Agriculturae 1 Nitra, Slovaca Universitas Agriculturae Nitriae, pp. 17–20, (2014)
24. *http://mapy.geoportal.gov.pl/imap/*

Monitoring of Inland Surface Water Quality Using Remote Sensing on the Example of Wigry Lake

Michał Lupa[1*], *Katarzyna* Adamek[1], *Renata* Stypień[1], and *Wojciech* Sarlej[1]

[1]AGH University of Science and Technology, Department of Geoinformatics and Applied Computer Science, Kraków, Poland

Abstract. The study examines how LANDSAT images can be used to monitor inland surface water quality effectively by using correlations between various indicators. Wigry lake (area 21.7 km^2) was selected for the study as an example. The study uses images acquired in the years 1990–2016. Analysis was performed on data from 35 months and seven water condition indicators were analyzed: turbidity, Secchi disc depth, Dissolved Organic Material (DOM), chlorophyll-a, Modified Normalized Difference Water Index (MNDWI), Normalized Difference Water Index (NDWI) and Normalized Difference Vegetation Index (NDVI). The analysis of results also took into consideration the main relationships described by the water circulation cycle. Based on the analysis of all indicators, clear trends describing a systematic improvement of water quality in Lake Wigry were observed.

1 Introduction

Water resources should be specially protected as they play a significant role in human life. Over time, anthropogenic factors have caused many changes in water quality [1]. The term 'water quality' is hard to define [2], but throughout this research it refers to the quality of the aquatic environment, as described by measured indicators. Water quality monitoring refers to the actual collection of data at regular intervals [3]. This makes it possible to detect trends and relations between them [2]. In environmental monitoring, classic methods (direct monitoring) can be described as measuring the value of an indicator in the field. In this case the absolute value of the measured parameter is obtained [4]. Indirect methods refer to calculating the value of a parameter on the basis of specific functional dependencies by measuring the value of other parameters [5].

Since the first Landsat satellite was launched, the data it has collected has been widely used to monitor water quality, for example by [6-9] and many others in recent years. In general, satellite images are popular in many studies about water quality due to the fact that they have features that ground samplings do not offer, including the ability to sense large areas with good time resolution; moreover, satellite images provide access to historical datasets [10].

The purpose of this research was multi-temporal analysis of indicators that can help to define water quality. Two hypotheses were put forward: first, water quality in this lake changes over a period of years; second, remote sensing methods are an effective tool to monitor these changes. As a test area, Wigry Lake was chosen with the assumption that the main source of water contamination are waters from surface runoff and those transported by the Czarna Hańcza river.

2 Study area

Wigry Lake is located in Podlaskie Voivodeship in northeastern Poland. With an area of 21.2 km^2 and a maximum depth of 73 m, it is an important part of Wigierski National Park, being its biggest lake [11]. Wigry Lake is a kettle lake with a few smaller lakes nearby, including Pietry, Dowcień, and Żubrowo, all of which used to be part of Wigry until they were separated from the main reservoir by mire processes. The main river flowing through the lake is Czarna Hańcza, with 1.4 m^3 per second water discharge. Characteristic of this river is the huge amount of dark silt that is transported to Wigry Lake [12].

Due to the unique character of the lake, many attempts have been made to protect its waters. In 1975, Wigry Lake was listed as one of world's most valuable inland water reservoirs and became part of the "Aqua" project, run by the International Union for the Conservation of Nature. In 1989 Wigierski National Park was created to protect the lake and its surroundings at the national level [13].

3 Data

The study uses LANDSAT mission images obtained from the USGS Earth Explorer website, acquired in the years 1990–2016. The plan was to analyze one image from every month within the range April–October for the years 1990–2016. Rigorous criteria were applied due to the necessity of fulfilling specific conditions such as lack

* Corresponding author: mlupa@agh.edu.pl

https://doi.org/10.1051/e3sconf/20186300017

of precipitation on the day of sensing and the day prior to it and lack of clouds at the location. Theoretically, for one scene per month in perfect weather conditions, 189 images would be obtained. Data that did not fulfill the strict requirements of this research were excluded from analysis. Finally, 35 images were analyzed.

All images were pre-processed to re-calculate spectral bands to radiance and reflectance values. Then, a python script was implemented to automate the repetitive work such as clipping images to the smallest possible rectangle describing the boundaries of Wigry Lake, calculating indexes, and averaging results. All operations were carried out in ArcMap, provided by the ArcGIS platform.

4 Methods

For the purposes of this research the following indicators were calculated:

- Turbidity (1) [14]. Referring to [15] turbidity can be described as the amount of light scattered by particles suspended in water. The greater the number of suspended particles, the greater the scattering. To determine turbidity remote sensing methods, bands with spectrums of red, near infrared and blue regions are usually used due to the highest radiation scattering by particles in these spectral ranges [16].

$$\text{Turbidity} = 63.9 - 1401.7R_{TM2} + 1827.8R_{TM3} - 78.6(R_{TM3}/R_{TM2}) \quad (1)$$

$$R^2 = 0.92$$

- Secchi disk visibility (2) [17], also described as "Secci disk depth" [18]. The *SSD* method applied in-situ is used to measure the clarity of water. A white disk is lowered until it is no longer visible to observers [19]. In remote sensing methods, blue, green and red light are used for this indicator; this determines the utility of Landsat TM to calculate value of *SSD* [10].

$$SSD = 208 \cdot \exp(-9.82(\varrho_{TM3}/\varrho_{TM2})) \quad (2)$$

$$R^2 = 0.85$$

- Dissolved Organic Matter (DOM) (3) [20] (also referred to as colored dissolved organic matter, or yellow substance) has the ability to absorb light. Therefore, remote sensing can be used to estimate the amount present in water [16].

$$DOM = 42.33 - 28.48\,(\varrho_{617.5\text{-}626.6nm}/\varrho_{1002.7\text{-}1035.5nm}) \quad (3)$$

$$R^2 = 0.82$$

- Chlorophyll-a (4) can be defined as an indicator of the abundance of phytoplankton [21]. As Chlorophyll-a absorbs blue and red radiation, remote sensing methods are used to measure chlorophyll-a content [16].

$$\text{Chl-a} = TM4/(TM1+TM2+TM4) \quad (4)$$

$$R^2 = 0.69$$

- Normalized difference water index (*NDWI*) (5) proposed by McFeeters [22] uses the green and the near infrared band. This allows water features with positive values to be distinguished. Others, such as vegetation or soil, are expected to have negative values or zero [22].

$$NDWI = (TM2-TM4)/(TM2+TM4) \quad (5)$$

- The Modified Normalized Difference Water Index (*MNDWI*) (6) was proposed by Xu [23] as a modification due to imperfections in *NDWI*. The near infrared band was replaced with a middle infrared band that gives improved results, such as higher values for water bodies; moreover, unlike *NDWI*, *MNDWI* does not distinguish built-up areas [23].

$$MNDWI = (TM2-TM5)/(TM2+TM5) \quad (6)$$

- Normalized Difference Vegetation Index (*NDVI*) (7) uses the near infrared band and the red band due to fact that the greener vegetation is, the more NIR light it reflects and absorbs less visible light. This allows the *NDVI* ratio to highlight vegetation with positive values [24].

$$NDVI = (TM4-TM3)/(TM4+TM3) \quad (7)$$

5 Discussion

Analyzing the graphs that show the variability of indexes over time shows a series of dependencies between them. However, to fully understand the changes taking place in the lake, it is necessary to turn to the process of water circulation.

Water circulation is related to the thermal stratification that occurs due to temperature variations at different depths of a lake [15]. During the summer, the upper layers of water are heated intensively, thus their density is reduced. Therefore, the waters in these upper and lower layers do not mix and the level of oxygen in the lower layers decreases. As the level of insolation declines, the temperature falls, which intensifies the strength of the wind. This leads to fall turnover: the water in all layers mix and the temperature and oxygen level equalize between them. During winter, temperature circulation stops and the top layer of the lake freezes. The layering of waters in the lake is distinct. The temperature equalizes when spring arrives and circulation restarts [25].The effects of thermal stratification can be noticed in the results of this study. When circulation stops, suspended matter is deposited on the bottom of the lake. When turnover occurs, sediments are lifted closer to the surface and the Secchi Disk Depth

https://doi.org/10.1051/e3sconf/20186300017

and turbidity indexes values therefore fall (Fig. 1 and Fig. 2). As the next conclusion, the DOM value increases due to

increased vegetation growth in the surroundings of the lake (Fig. 3). This leads to a rise in humid acids, which causes the SSD index to start rising when Spring arrives. At the end of July, the Chl-a value starts to fall (Fig. 4). This is related to reduced production of algae.

All the indexes rise across the studied years. DOM and SSD increase systematically (Fig. 5 and Fig. 6), thus implying that water quality in Wigry lake is getting better. This suggests that actions taken to protect the lake, whose main goal is to limit the amount of dissolved substances deposited in the lake, are effective. This can be attributed to the efficiency of sewage treatment in the city of Suwałki, which was modernized in the late 1990s. Moreover, more responsible agriculture has reduced the use of pesticides, which reduces the amount of mineral and organic substances in the water. Due to this, water fertilization has decreased.

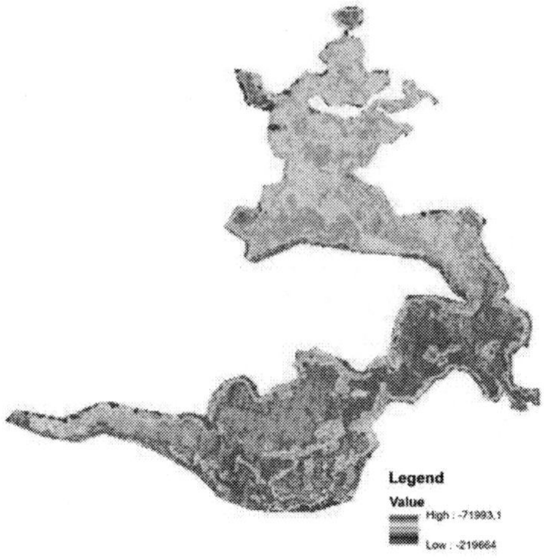

Fig. 1. Turbidity index in July 2010.

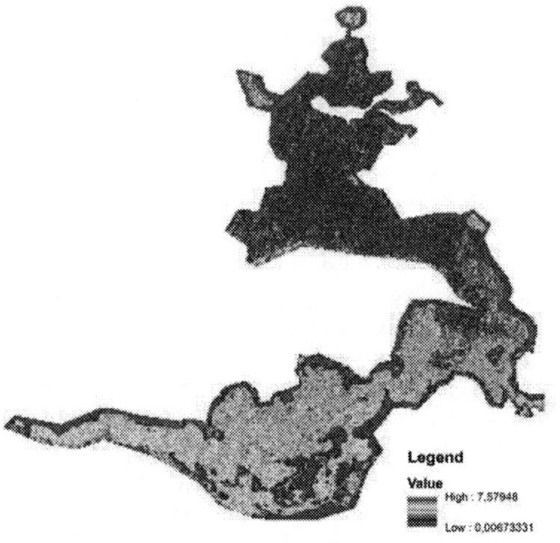

Fig. 2. SSD index in July 2010.

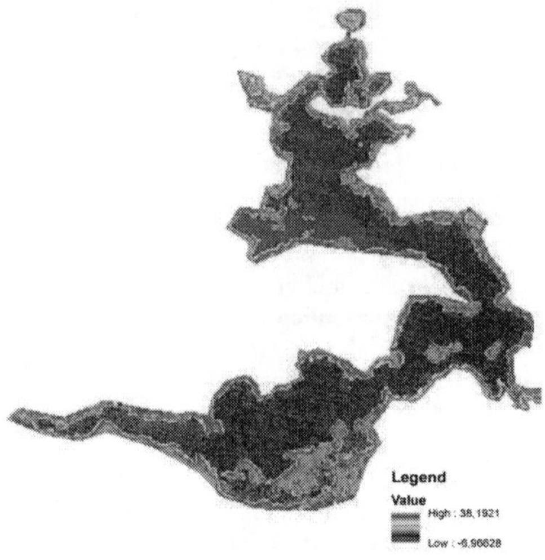

Fig. 3. DOM index in July 2010.

6 Conclusions

This research proves that multitemporal analysis with the use of remote sensing methods allows waters to be monitored indirectly. Water clarity together and other parameters describing water quality can be evaluated. The possible correlation of results obtained with remote sensing with in-situ testing would maximize the efficiency of future indirect monitoring. The research shows that remote sensing is a suitable tool to monitor the quality of Wigry Lake water. Moreover, the effectiveness of the actions taken to protect the lake has been confirmed.

Research were conducted within founds of Faculty of Geology, Geophysics and Environmental Protection grant (AGH University of Science and Technology) no. 11.11.140.613.

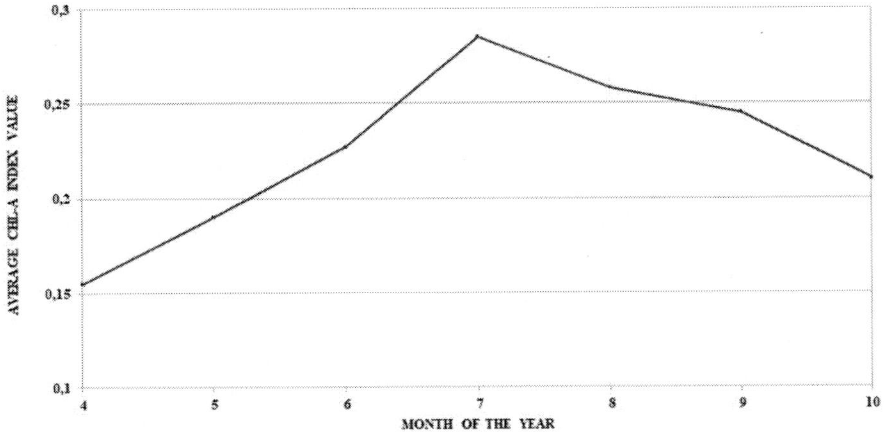

Fig. 4. Average Chl-a index value (y axis) for every analysed month (x axis).

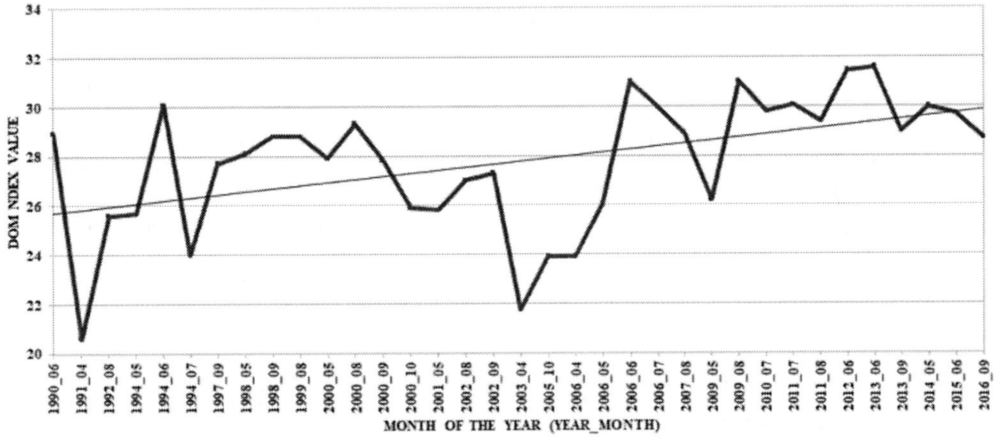

Fig. 5. DOM index value (y axis) for the whole period of research (x axis).

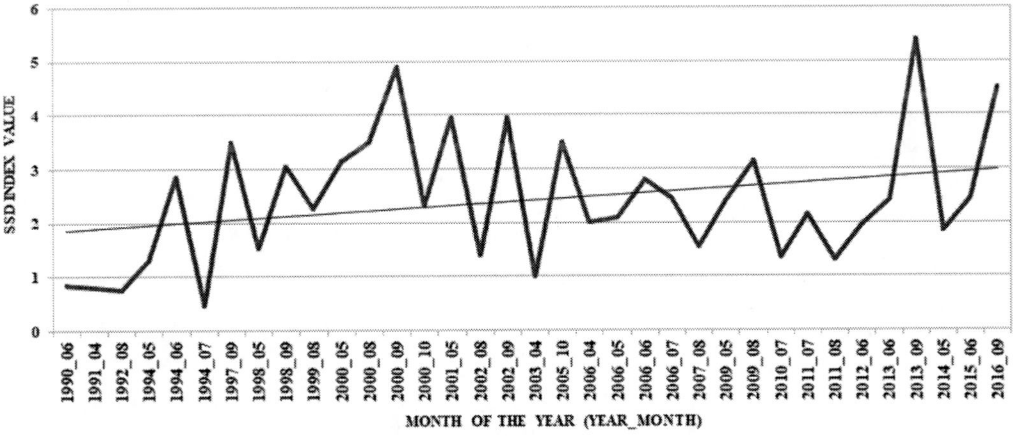

. **Fig. 6.** SSD index value (y axis) for the whole period of research (x axis).

References

1. I. A. Shiklomanov, *World Water Resources, A new appraisal and assessment for the 21st century*, UNESCO (1998)
2. D. Chapman, et. al., *Water Quality Assessments - A Guide to Use of Biota, Sediments and Water in Environmental Monitoring – Second Edition* London (1996)
3. J. Bartram and R. Balance, *Water Quality Monitoring: A Practical Guide to the Design and Implementation of Freshwater Quality Studies and Monitoring Programmes*, London (1996)
4. T. Dworak, B. Hejmanowska, and K. Pyka, *Problemy teledetekcyjnego monitoringu środowiska. Teledetekcja wód i powierzchni ziem*, Wydawnictwo AGH, Krakow, (2011) (in Polish)
5. A. Fedoryszyn and E. Ziółkowski, *Odlewnictwo XXI wieku – technologie, maszyny i urządzenia odlewnicze*, pp. 103-111 (2003)
6. A. Kulkarni, Procedia Computer Science, **6** (2011)
7. R.G. Lathrop, Photogrammetric Engineering and Remote Sensing, **58** (1992)
8. L.G. Olmanson, M.E. Bauer, and P.L. Brezonik, Remote Sensing of Environment, **112** (2008)
9. S.M.J. Baban, International Journal of Remote Sensing, **14** (1993)
10. F.L. Hellweger, P. Schlosser, U. Lall, and J.K. Weissel, Estuarine, Coastal and Shelf Science, **61** (2004)
11. J. Jańczak, Atlas jezior Polski, Volume III (1999)
12. L. Magrel, M. Ładyński, U. Ikowska-Ładyńska, and P. Herman, *Program ochrony środowiska do 2012 roku. Gmina Suwałki* (2004) (in Polish)
13. Polish National Parks. [Online]. Available: http://www.parkinarodowe.edu.pl/pn/wigierski_pn.htm
14. K. Sørensen, E. Aas, B. Faafeng, and T. Lindell, *Fjernmåling av vannkvalitet - Videreutvikling av optisk satellittfjernmåling som metode for overvåking av vannkvalitet*, (1993)
15. J. P. Michaud, *A Citizens' Guide to Understanding and Monitoring Lakes and Streams* (1991)
16. K. Osińska-Skotak, P. Hydzik and P. Walawender, *Badania przestrzeni kosmicznej a innowacje i wzrost gospodarczy*, pp. 84-108 (2014) (in Polish)
17. R.G. Lathrop, T.M.Illesand, and B.S.Yandell, International Journal of Remote Sensing, **12**, (1991)
18. D. Kar, *Epizootic Ulcerative Fish Disease Syndrome*, pp. 187-221, Elsevier (2016)
19. S. Q. Duntley, *The Visibility of Submerged Objects* (1960)
20. A. Polvorinos, et al., *Remote sensing modelling and monitoring of water quality in Aracena and gergal dams (Seville, Spain)*, in Proc. of the 4rd CHRIS/PROBA Workshop, (2006)
21. S. Poikāne, et al., Environmental Management, **45.6** (2010)
22. S.K. McFeeters, International Journal of Remote Sensing, **17** (1996)
23. H. Xu, International Journal of Remote Sensing, **27** (2005)
24. J.W. Rouse, R.H. Haas, J.A. Schell, and D.W. Deering, *NASA. Goddard Space Flight Center 3d ERTS-1 Symp.*, **1** (1974)
25. C. Brothers, D.A. Culver, and R. W. Fortner, *Lake Layers: Strarification* (The Ohio State University, 1991)

* Corresponding author: mlupa@agh.edu.pl

Evaluation of the quality of cadastral data

Jarosław Taszakowski[1*], *Urszula* Litwin[1], *Arkadiusz* Doroż[1] and *Karolina* Misiak[1]

[1]Department of Agricultural Land Surveying, Cadastre and Photogrammetry, University of Agriculture in Krakow, Balicka 253a, 30-198 Krakow, Poland

Abstract. The purpose of the work is to determine the possibility of assessing the quality of cadastral data basing on the system of land and building registry. The study of discrepancies between the descriptive and graphical parts of the of the land and buildings registry was accepted as the scope of research in the field of marking plots and their areas, checking the level of: supplementing registry data with PESEL numbers for private entities and REGON numbers for public entities, completeness of addresses data, verification of documents defining the legal status of real estates and data on border points. The research objects were 14 precincts, including 11 rural ones in the commune of Dąbrowa Tarnowska and 3 in the urban commune of Dąbrowa Tarnowska, located in the Dąbrowski district in the Małopolskie voivodship.

1 Introduction

Cadastre is an information system that provides collection, updating and sharing information on land, buildings and premises, their owners and any other entities owning or managing the land, buildings or premises [1,2].

The Polish cadastre developed in stages similarly to other cadastral systems in the world [3,4]. As a first stage, the inventory of documentation related to the Austrian and Prussian cadastre can be considered, as well as the geodetic and cartographic documentation, which was created in connection with consolidation works in the interwar period as well as documentation produced during the socio-political changes in the post-war years. The next stage was the establishing of a uniform land registry in the whole country. The third stage was and is the computerization of the land and building registry system [5].

The basic task assigned to the real estate cadastre is to collect numerical and descriptive data about land, buildings and premises. It is run by bodies that have a public character. It has rules for its creation, updating and sharing created in accordance with applicable regulations [6].

The most important development trend of the real estate cadastre in Poland is creation of the Integrated Property Information System (IPIS). The legal basis for the functioning of this system is determined by the provisions of the Geodetic and Cartographic Law Act and the provisions of the Regulation of the Council of Ministers of January 17, 2013 regarding the Integrated Cadastral System. [7]. However, among the global development trends of the cadastre, the creation and implementation of a multi-dimensional cadastre should be taken into account [8-11]

The concept of creation of the Integrated Property Information System, was developed in 1999. As a result of the arrangements between the Polish government and the European Commission, the PHARE 2000 program was created - "Creation of the Integrated Cadastral System". It was a joint project of the Main Office of Geodesy and Cartography, the Ministry of Finance and the Ministry of Justice. The idea of the program was to provide access to up-to-date and reliable information about real estates for public administration bodies, entrepreneurs and citizens using electronic technology. [12].

The aim of running the IPIS is to improve the functioning of the state's information system regarding real estates, by creating solutions to ensure data integrity and timeliness (in accordance with the data quality model presented in Figure 1), regarding real estates included in public registers, automation of their update processes, as well as the widest possible use of information collected in these registers for public purposes, taking into account the state of computerization of these registers [13].

The basis of IPIS is the land and buildings registry, which proper functioning is conditioned by the fulfilment of the requirements specified in art. 21 of the Geodetic and Cartographic Law of Purposes and provision of reference data to other systems.

The Integrated Property Information System as a teleinformation system should enable keeping the central repository of copies. It is connected with the update and secure storage of information forming one consistent database. [14]

* Corresponding author: jaroslaw.taszakowski@urk.edu.pl

The lack of consistency and coherence of data between the land and building registry, land and mortgage registers as well as the abovementioned public registers is still a current and a big problem.. That is why the problem of assessing the functioning of land and building registry as the most important element of IPIS becomes current.

Fig. 1. Data quality model, source: own study.

2 Description of the research methodology

The research is aimed at assessing the quality of cadastral data in the context of the implementation of the Integrated Property Information System. The analyses which were carried out refer to the data quality model which was made available by the Head Office of Geodesy and Cartography, namely: logical consistency, accuracy (temporal, thematic and position), usability and completeness. The quality assessment was based on 4 analyses:

- the analysis of discrepancies in registry data between the descriptive and graphical parts,
- the analysis of legal conditions shown in the land and building registry,
- the analysis of completeness of the registry database in the field of:
  - PESEL and REGON attributes,
  - residential and institutional addresses.
- the analysis of the accuracy of cadastral databases in the field:
  - sources of data on the location of border points,
  - stabilization code,
  - border order code,
  - the mean error of the position of the border point with respect to the first class network.

Basing on the analysis of the source of data on the location of border points and the mean error of the position of the border point in relation to the I-class network, it was checked how many parcels have a legal border and how many parcels have border points with the determination error 0.00 - 0.30 m.

* Corresponding author: jaroslaw.taszakowski@urk.edu.pl

In order to illustrate the quality of cadastral data in the examined facilities, a ranking of blocks has been prepared, taking into account the quality of the possessed data. For this purpose, the following evaluation criteria and established weights were selected:

- no discrepancies between the graphic part and the descriptive part regarding the number of registered plots (weight 1),
- no differences in the area between the graphic and descriptive part (weight 2),
- parcel area determined with the accuracy of 0.0001 ha (weight 1),
- land and mortgage register disclosed in the database (in the current and electronic form) (weight 2),
- the PESEL or REGON attribute shown in the database (weight 1),
- address of the owner in the case of natural persons or company headquarters with reference to the institutions shown in the database (weight 1),
- each border point is legally approved (weight 2),
- the position error of each border point is in the range 0.00 - 0.30m (weight 3).

Next, the coefficient was calculated, which In a very approximate but easy to interpret way determines the completeness, coherence and accuracy of cadastral data in the precincts with reference to both registry units (formula 1).

$$W_k = \frac{x_i - x_{min}}{x_{max} - x_{min}} \quad (1)$$

where:
W_k – the quality factor of cadastral data in the registry unit
x_i – the analysed value
x_{max} – maximum value
x_{min} – minimum value.

3 The research results

Below some graphic elaborations will be presented as the research results. They are included in the figures no. 2-12.

In the first analysis, shown in figure 2, the audit was carried out in the SCAL-SWDE program, it consisted in determining the differences in the identification of the parcels on the map and in the descriptive part of the survey. Then, the areas of registration plots from the evidential database and counted areas based on the vector map were compared. The settlement of parcel areas on the cadastral map was made in the Bentley Microstation V8i program using the MK Power 2015 overlay.

 https://doi.org/10.1051/e3sconf/20186300018

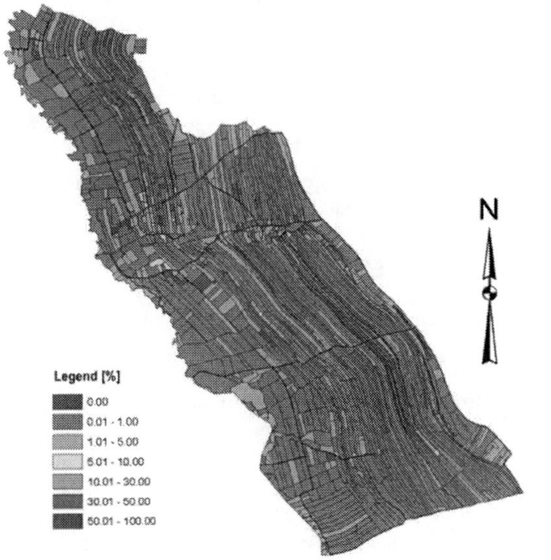

Fig. 2. Discrepancies in the registry data between the descriptive part and the graphical part regarding the compliance of the area of parcels on the example of the Brnik precinct, source: own study.

The analysis of the accuracy of the determination of the surface is aimed at finding plots of land whose surface is not specified in accordance with the guidelines provided for in Polish legal norms, i.e. up to 0.0001 ha. The urban area has 100% plots of land whose area is specified up to 0.0001 ha. The situation in the rural area deviates from legal norms, since most properties have a field of only 0.01 ha, which is a consequence of the old legal requirements, allowing the accuracy of determining the area to 0.01 ha. The attached figure shows that the accuracy norms are met only for plots that were subject to geodetic and legal activities and had newly calculated areas.

In the next part of the research, the cadastral databases were analysed in terms of their completeness, accuracy and usability. In the case of legal status of real estate visible in the database of registers, the land and mortgage register is a basic document confirming the right to real estate (over 75% of all documents). There is a large number of plots where there is no document stating the legal status (about 15% of the total). The study was carried out in the ESRI ArcGIS program.

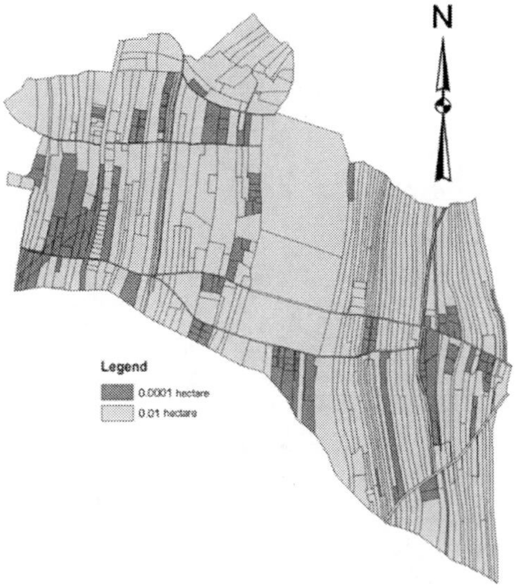

Fig. 3. Analysis of the accuracy of cadastral databases in terms of the accuracy of area determination on the example of the evidence precinct Gruszów Mały, source: own study.

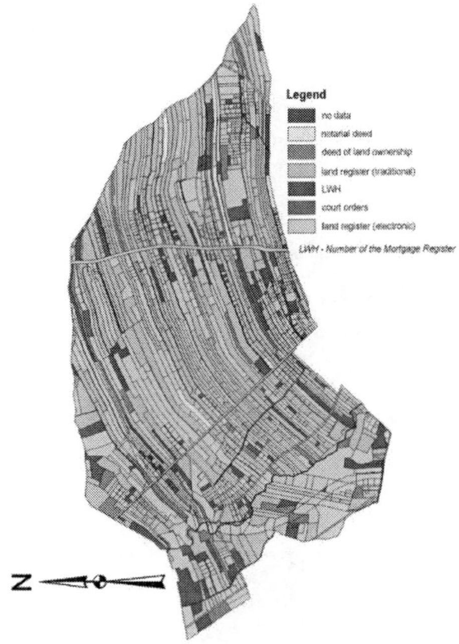

Fig. 4. Analysis of legal conditions within Ruda precinct, source: own study.

The property cadastre shows a lot of data regarding land, buildings, premises and their owners. One of such information is the place of permanent residence or the address of the registered office of the entities. The disclosure of the address in the property cadastre is very important, it allows for correct notification of the site during legal and geodetic works. In the studied area, permanent residence addresses are given to natural persons in more than 98%, and the headquarters of the institutions only in 76.6%.

* Corresponding author: jaroslaw.taszakowski@urk.edu.pl

Fig. 5. Analysis of the completeness of the cadastral database in the scope of supplementing the addresses of residences and headquarters of institutions - Morzychna, source: own study.

Fig. 6. Analysis of the completeness of the registry database with regard to supplementing the PESEL and REGON attributes on the example of Lipiny, source: own study.

The database was analysed in terms of disclosing the PESEL and REGON attributes. There is no registration area that has a complete set of data about these registers. In both cases, a larger number of plots has a PESEL or REGON disclosed.

The first analysis concerning attributes of border points concerned the source of data on the location of the border point. The following values of the ZRD attribute are in Polish standards: Attribute 1- geodetic field measurements preceded by delimitation of real estate, resumption of border marks, determination of border points or determination of their location in a different mode:

- Attribute 2 - geodetic field measurements not preceded by delimitation of real estate, resumption of border marks, determination of border points or determination of their location in a different mode.
- Attribute 3 - geodetic photogrammetric measurements of border points whose location has been predetermined in a specified manner, as well as photogrammetric measurements of boundary marks displayed on aerial photographs or orthophoto maps. as a result of their signalling before taking pictures.
- Attribute 4 - geodetic photogrammetric measurements not preceded by determining the boundaries of registration parcels or signalling border signs before taking aerial photographs.
- Attribute 5 - approved projects for the division of real estate or the consolidation and division of real estate.
- Attribute 6 - approved projects for the consolidation or exchange of land.
- Attribute 7 - screen vectorization of the raster map using unified use of geodetic measurements of terrain measurements (linear measures).
- Attribute 8 -screen vectorization of the record raster map without using geodetic measurements of field measurements.
- Attribute 9 - other data sources.

It shows that over 80% of border points are established on the basis of photogrammetric measurements not predicted by the record boundaries or on-screen vectorization of the raster map using the results of geodetic field measurements.

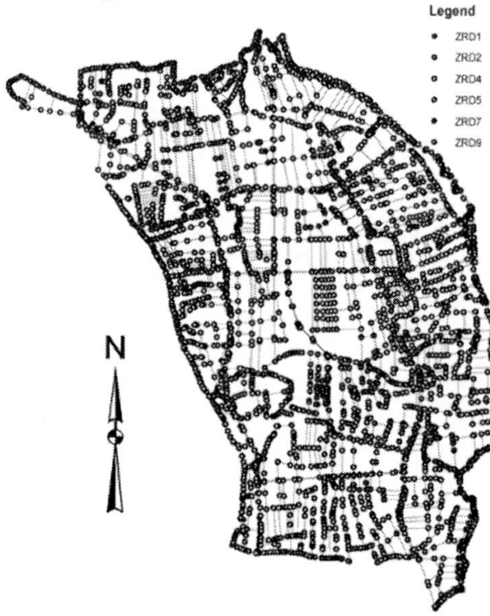

Fig. 7. Values of the data source for the location of the border point within the evidence precinct Sutków, source: own study.

* Corresponding author: jaroslaw.taszakowski@urk.edu.pl

The data on border points was also analysed in terms of border code and it was established that the area under research "has points on the government: register plot (93.0%), registration area (3.8%), register unit (1.2%), municipality (1.4%)) and poviat (0.6%).

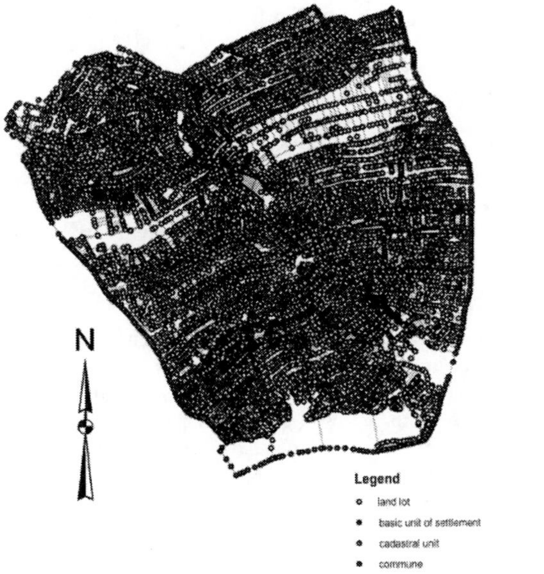

Fig. 8. The values of the code on the border order in Szarwark, source: own study.

The third analysis checked the method of stabilizing border points. It shows that only 3.4% of border points have stabilization. The rest of the points are either not stabilized (52.8%) or there is no information on this subject (43.8%).

Fig. 9. The values of the border point stabilization code in Żelazówka, source: own study.

* Corresponding author: jaroslaw.taszakowski@urk.edu.pl

Another analysis was the analysis of the accuracy of cadastral databases in terms of the average position error of the boundary point in relation to the first-class geodetic control network. The most points are with the accuracy of 1.51m to 3.00m (32.3%) and 0.00m to 0.10m (14.0%). Once again, the largest problem of the analysed database was revealed, i.e. large gaps, as many as 50.5% of border points have no specific location error.

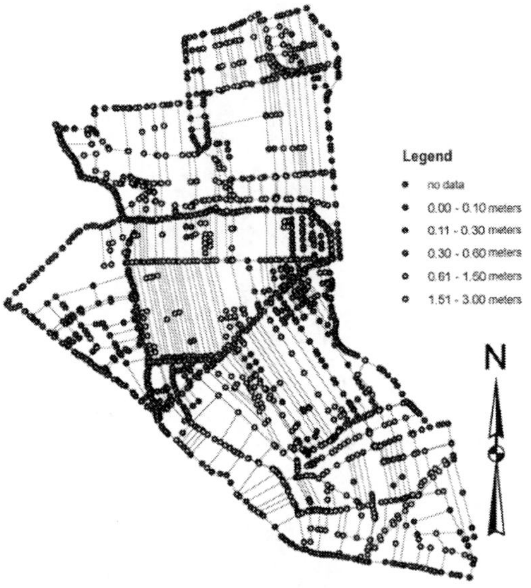

Fig. 10. Average error of the border point relative to the geodetic control network of the 1st class within Morzychna precinct, source: own study.

One of the most interesting analyses in terms of surveying was to check how many plots have all legal boundaries. For this purpose, GIS tools were used, specifically the QGIS program, whose tools allowed to determine the percentage share of points defined legally in relation to all. As a result of the work, it turned out that 10.8% of registered plots in the area of the city and the commune of Dąbrowa Tarnowska have the entire legal boundary.

Another analysis, also made with the use of GIS tools, concerned the representation of the percentage share of border points determined with an accuracy of 0 \ 0 to 0.30 m in relation to all. As for the accuracy of determining the border points, 11.2% of plots have a set of such border points, where the position error in terms of the geodetic control network of the 1st class is up to 0.30 m.

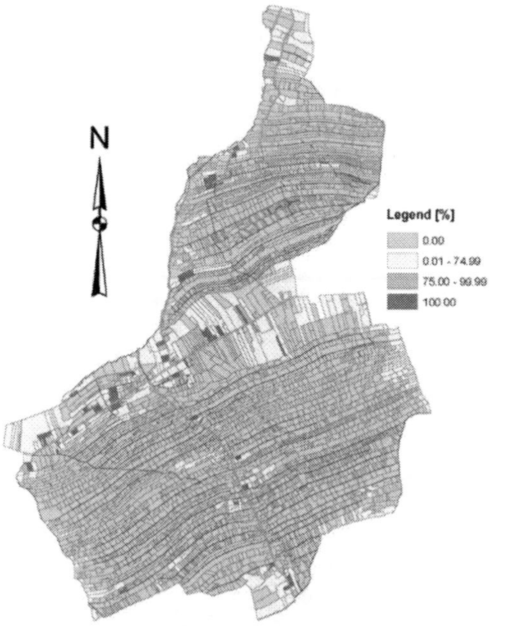

Fig. 11. The legal boundary of the registry parcels on the example of the Nieczajna precinct, source: [7].

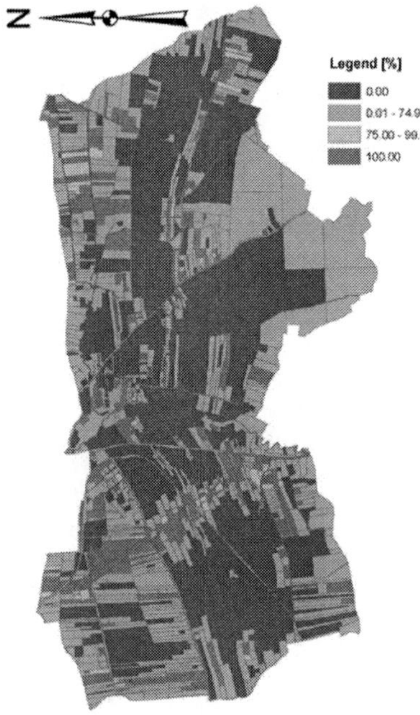

Fig. 12. Analysis of parcel plots with regard to the error of the average location of the border point with respect to the geodetic control network of the 1st class on the example of the Bagienica precinct, source: own study.

In order to illustrate the situation regarding the quality of cadastral data in the city and commune of Dąbrowa Tarnowska, a ranking list has been prepared on the basis of the above-mentioned surveys, due to the quality of the data held.

For this purpose, the percentage share of plots fulfilling the given criterion is compared to the total number of registered plots within. Selected criteria are in sequence: no discrepancies between the graphic part and the descriptive part in the number of parcels, no differences in area between the graphic and descriptive part, surface area specified up to 0.0001 ha, disclosed in the land and mortgage register (in the existing and electronic form), displayed in the PESEL or REGON attribute database, displayed in the owner's address database in the case of individuals or registered office with regard to institutions, each border point is legally valid and the position error of each border point is in the range 0.00 to 0.30 m.

The list presented below shows the state of cadastral data only in relation to the studied area (2 registry units). In comparison with the city and commune of Dąbrowa Tarnowska, the best quality of cadastral data is provided by the following divisions: Dąbrowa, Ruda and Bagienica.

Table 1. Calculation of the cadastral data quality factor (source: own study).

Place in the ranking	Precinct	evaluation criterion number:								SUM	W_k
		1	2	3	4	5	6	7	8		
		weight of the criterion:									
		1.0	2.0	1.0	2.0	1.0	1.0	2.0	3.0		
1	Dąbrowa	100.0	25.8	100.0	84.8	83.7	95.5	27.7	26.1	**734.1**	100.0
2	Ruda	100.0	30.6	100.0	86.6	83.7	92.1	22.9	23.0	**725.0**	97.0
3	Bagienica	100.0	21.1	100.0	79.5	74.6	90.6	16.9	18.7	**656.3**	74.8
4	Morzychna	100.0	21.8	30.3	77.1	52.5	97.2	21.8	23.1	**590.7**	53.5
5	Gruszów Mały	99.1	17.0	29.8	82.2	73.4	93.4	11.2	16.9	**567.2**	45.9
6	Łaskówka Chorąska	99.9	7.3	13.1	69.9	84.7	98.4	11.6	12.0	**509.7**	27.2
7	Żelazówka	100.0	6.7	11.5	70.4	86.0	98.7	9.5	9.4	**497.6**	23.3
8	Gruszów Wielki	100.0	6.5	13.6	72.7	79.2	97.6	7.1	7.4	**485.2**	19.3
9	Lipiny	100.0	3.4	12.9	73.6	81.6	95.0	7.3	7.2	**479.7**	17.5
10	Smęgorzów	99.9	3.7	6.2	77.5	79.1	94.8	4.3	5.6	**467.8**	13.6
11	Sutków	100.0	4.8	11.3	64.7	80.3	97.2	7.7	7.8	**466.6**	13.2
12	Szarwark	100.0	2.5	4.0	76.7	82.1	94.7	2.6	2.9	**453.1**	8.9
13	Brnik	100.0	4.2	7.9	66.0	75.4	96.8	6.4	6.1	**451.6**	8.4
14	Nieczajna	99.8	2.2	3.5	74.6	59.9	98.6	2.0	2.1	**425.7**	0.0
MAX		**100.0**	**30.6**	**100.0**	**86.6**	**86.0**	**98.7**	**27.7**	**26.1**	**734.1**	**100.0**
MIN		**99.1**	**2.2**	**3.5**	**64.7**	**52.5**	**90.6**	**2.0**	**2.1**	**425.7**	**0.0**
AVERAGE		**99.9**	**11.3**	**31.7**	**75.5**	**76.9**	**95.8**	**11.4**	**12.0**	**536.5**	**35.9**

Taking into account the carried out research, it has been established that all areas have smaller or larger gaps in the cadastral database. Therefore, the task of the presented classification is to help in defining the schedule of modernization of land and building registry. First of all the attention should be paid to the rural areas, i.e. Nieczajna, Brnik, Szarwark, Sutków, Smęgorzów and Lipiny.

Taking into account that the precincts could have obtained a total of 1300 points, they are summarized in Table 2 of the scope with reference to the reference condition. This list clearly indicates that the data in the analysed area are not complete and accurate.

* Corresponding author: jaroslaw.taszakowski@urk.edu.pl

https://doi.org/10.1051/e3sconf/20186300018

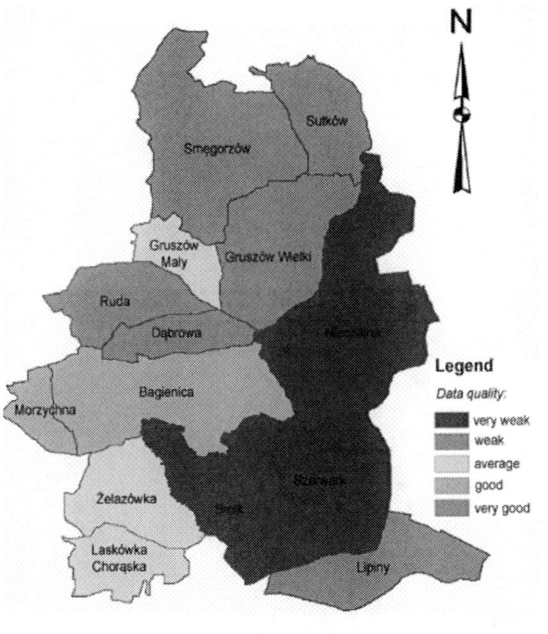

Fig. 13. The quality of cadastral data in relation to the city and commune of Dąbrowa Tarnowska, source: own study.

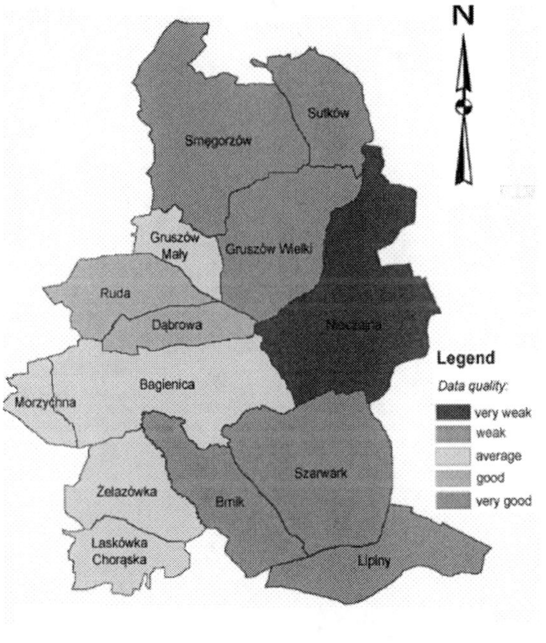

Fig. 14. The quality of cadastral data in relation to the master status. source: own study

Table 2. Quality factor of cadastral data with reference to the master status (source: own study).

Place in the ranking	Precinct	evaluation criterion numer:								SUM	$W_{[\%]}$
		1	2	3	4	5	6	7	8		
		weight of the criterion:									
		1.0	2.0	1.0	2.0	1.0	1.0	2.0	3.0		
1	Dąbrowa	100.0	25.8	100.0	84.8	83.7	95.5	27.7	26.1	**734.1**	56%
2	Ruda	100.0	30.6	100.0	86.6	83.7	92.1	22.9	23.0	**725.0**	56%
3	Bagienica	100.0	21.1	100.0	79.5	74.6	90.6	16.9	18.7	**656.3**	50%
4	Morzychna	100.0	21.8	30.3	77.1	52.5	97.2	21.8	23.1	**590.7**	45%
5	Gruszów Mały	99.1	17.0	29.8	82.2	73.4	93.4	11.2	16.9	**567.2**	44%
6	Laskówka Chorąska	99.9	7.3	13.1	69.9	84.7	98.4	11.6	12.0	**509.7**	39%
7	Żelazówka	100.0	6.7	11.5	70.4	86.0	98.7	9.5	9.4	**497.6**	38%
8	Gruszów Wielki	100.0	6.5	13.6	72.7	79.2	97.6	7.1	7.4	**485.2**	37%
9	Lipiny	100.0	3.4	12.9	73.6	81.6	95.0	7.3	7.2	**479.7**	37%
10	Smęgorzów	99.9	3.7	6.2	77.5	79.1	94.8	4.3	5.6	**467.8**	36%
11	Sutków	100.0	4.8	11.3	64.7	80.3	97.2	7.7	7.8	**466.6**	36%
12	Szarwark	100.0	2.5	4.0	76.7	82.1	94.7	2.6	2.9	**453.1**	35%
13	Brnik	100.0	4.2	7.9	66.0	75.4	96.8	6.4	6.1	**451.6**	35%
14	Nieczajna	99.8	2.2	3.5	74.6	59.9	98.6	2.0	2.1	**425.7**	33%
MAX		**100.0**	**30.6**	**100.0**	**86.6**	**86.0**	**98.7**	**27.7**	**26.1**	**734.1**	**56%**
MIN		**99.1**	**2.2**	**3.5**	**64.7**	**52.5**	**90.6**	**2.0**	**2.1**	**425.7**	**33%**
AVERAGE		**99.9**	**11.3**	**31.7**	**75.5**	**76.9**	**95.8**	**11.4**	**12.0**	**536.5**	**41%**

4 Summary

On the basis of the conducted research, it can be concluded that the cadastral database in the area of the city and commune of Dąbrowa Tarnowska is not coherent at the moment. Data from the graphic part do not coincide at 100% with the data of the descriptive part. There are a lot of differences that have enormous significance in civil law proceedings. The register under investigation is also not a complete or accurate database, in many cases, the information was insufficient (for example, for addresses), instead of reliable information, also the message "no data in swde" appeared. Cadastral data in the Dąbrowa centre should be considered as up-to-date. Data modifications are quite frequent and mainly concern the last three years.

At present, basing on the research and analysis carried out, it should be stated that the city and commune of Dąbrowa Tarnowska are not ready for the implementation of the Integrated Cadastre System. The collected cadastral data constitute a huge amount of information, but do not meet the necessary requirements for the correct operation of IPIS.

References

1. O. Demir, and Y.E Çoruhlu, *Determining the property ownership on cadastral works in Turkey*. Land Use Policy, **26**, pp. 112–120 (2009)
2. M. Siejka, M. Ślusarski, and M. Zygmunt, *3D+time Cadastre, possibility of implementation in Poland*. Survey Review, **46**(335), pp. 79–89 (2014)
3. A.C. Aydinoglu, R. Bovkir, *Generic land registry and cadastre data model supporting interoperability based on international standards for Turkey*. Land Use Policy, **68**, 59–71 (2017)
4. D.Steudler, A.Rajabifard, I.P.Williamson, *Evaluation of Land administration systems*. Land Use Policy **21**(4), 371–380, (2004)
5. M. Mika, *An Analysis of Possibilities for the Establishment of a Multipurpose and Multidimensional Cadastre in Poland*. Land Use Policy, 77: 446–53 (2018)
6. K. Kocur-Bera. *Data compatibility between the Land and Building Cadaster (LBC) and the Land Parcel Identification System (LPIS) in the context of area-based payments: A case*

* Corresponding author: jaroslaw.taszakowski@urk.edu.pl

study in the Polish Region of Warmia and Mazury. Land Use Policy, 65, 143–53 (2018)

7. Coucil of Ministers' decree of January 17, 2013 on the integrated property information system (Journal of Laws 2013 item 249)
8. M. Benhamu, Y Doytsher, *Toward a spatial 3D cadastre in Israel*, Computers, Environment and Urban Systems, **27**, 4, 359-374 (2003)
9. K.L. Kim, S.W. Hwang Bo, G.H. Lim, and S.J. Park, *A Study on the Development Cadastral Model for 3D Cadastre Registration, Journal of Korea Society of Surveying*, Geodesy, Photogrammetry, and Cartography, **23**, 2, 129-136 (2005)
10. P. Molen, *Institutional aspects of 3D cadastres*, Computers, Environment and Urban Systems, **27**, 4, 383 -394 (2003)
11. J. Stotoer, P. Oosterom, *3D Cadastre in an International Context: Legal, Organizational, and Technological Aspects*, Taylor & Francis, New York (2006)
12. E. Gołąbeska, *The Integrated Cadastral System as a factor of real estate market development*. Real estate world 2 (76) (2011) (in Polish)
13. http://www.gugik.gov.pl/geodezja-i-kartografia
14. P. Pochwatka, T. Teterycz, U. Litwin, and A. Bitner, *Cartographic Visualization in The Real Estate Market Investigation with the Use of GIS Tools*, 2017 Baltic Geodetic Congress (Geomatics) pp. 105–109 (2017)

* Corresponding author: jaroslaw.taszakowski@urk.edu.pl

Archaeological monuments of the Warmian-Masurian voivodeship – spatial analysis using GIS tools

Krzysztof Rząsa[1,*], and *Mateusz* Ciski[1]

[1]Faculty of Geodesy, Geospatial and Civil Engineering, Institute of Geography and Land Management, University of Warmia and Mazury in Olsztyn, Prawochenskiego Street 15, 10-720 Olsztyn, Poland

Abstract. Archaeological monuments are one of the forms of monument protection in Poland. It includes various objects, among which one can distinguish: field remnants of prehistoric and historical settlements, graveyards, burial mounds and single graves, as well as relics of industrial, religious and artistic activity. All these objects are of great importance in spatial planning, and its location possess a big impact on the possibilities of land development. According to The National Heritage Board of Poland database for September 2018, there are 7 672 archaeological monuments registered in the polish register of objects of cultural heritage. The number of archaeological objects entered in the register in individual voivodeships is varied. The majority of monuments is located in the Lower Silesian voivodeship, the least in Łódź voivodeship. In this article, archaeological monuments in the Warmian-Masurian voivodeship were analysed, in which there are 286 of such objects. Identification with regard to function was carried out, and a detailed location was specified. With the use of GIS tools, a series of maps was created, which after analysing became the basis for formulating final conclusions. Some of it can be used in the preparation of planning documents at the voivodeship and municipal level.

1 Introduction

The Act on the Protection of Monuments and the Care of Historical Monuments [1] introduces in Article 7 the following forms of protection of monuments in Poland: an entry in the register of objects of cultural heritage (*pol. rejestr zabytków*), an entry on the List of Heritage Treasures (*pol. Lista Skarbów Dziedzictwa*), recognition of an object or area as a Historic Monument (*pol. pomnik historii*), establishing a Cultural Park (*pol. park kulturowy*) and protection regulations in spatial planning documents. Archaeological monuments are among the objects entered in the register of objects of cultural heritage. According to Article 3 of the Act, an archaeological monument is a non-movable monument, a surface, underground or underwater remnant of human existence and activity, composed of cultural layering and the products or traces found therein, or a movable monument that is this product. Archaeological monuments are in particular: field remnants of prehistoric and historical settlements, graveyards, burial mounds and single graves, as well as relics of industrial, religious and artistic activity. Archaeological monuments are inscribed in Book C of the register of objects of cultural heritage.

Such objects are an element of cultural heritage and are subject to the principles of protection characteristic of this type of objects. Such activities can be observed all over the world [2-4]. The cultural heritage is the wealth, which connects people with the past, gives a solid foundation for the present and gives the way to the future [5].

The location of archaeological monuments, as well as other non-movable monuments and Cultural Parks or Historic Monuments, has a large impact on the development of land, especially in their immediate vicinity and must be taken into account in spatial planning [6-8]. Archaeological sites, as highly complex parts of the cultural heritage, require special care in all parts of planning, research, conservation and presentation [9].

This article is a detailed analysis of archaeological monuments located in the Warmian-Masurian Voivodeship. Archaeological monuments were identified, their functions were analysed, the exact location was determined, to finally make detailed analyses on the maps generated using the GIS tools. All this allowed to formulate the conclusions presented in the last part of the article.

2 Materials and methods

In the first stage of the study, an analysis of existing legal provisions and literature thematically related to the analysed topic was conducted. In this way, the main principles of the functioning of the monument protection system in Poland were determined, with particular reference to archaeological monuments.

In the next stage, the focus was put on a detailed inventory of archaeological monuments in Poland, in particular in the Warmian-Masurian Voivodeship.

* Corresponding author: krzysztof.rzasa@uwm.edu.pl

Quantitative and generic data was obtained from the National Heritage Board of Poland [10]. In Poland, 7 672 archaeological monuments are entered into the Book of C of the voivodeship register of objects of cultural heritage. The number of archaeological monuments entered in the register in individual voivodeships is varied. Majority of such monuments is located in the Lower Silesian voivodeship (19.19%), Opole voivodeship (15.46%), Greater Poland voivodeship (8.68%), Pomeranian voivodeship (7.38%) and Lubusz voivodeship (7.09%); the minority in the Łódź voivodeship (1.86%), Lublin voivodeship (2.20%) and Kuyavian-Pomeranian voivodeship (2.49%). In the analysed Warmian-Masurian voivodeship there are 3.73% of this type of objects. The detailed number of archaeological monuments in individual voivodeships is shown in Fig.1.

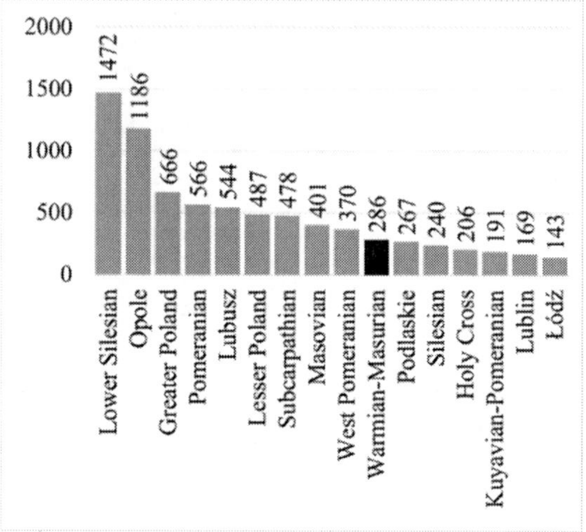

Fig. 1. The number of archaeological monuments, divided into polish voivodeships.

Regional disparities are even more visible when the number of monuments is related to the area of voivodeships. The highest ratio occurs for the Opole voivodeship (126 monuments per 1000 km2 area) and Lower Silesia voivodeship (74), and the lowest for the Lublin voivodeship (7) and Łódź voivodeship (8), with the national average of 25 monuments per 1000 km^2 area. The Warmian-Masurian voivodeship is below the national average (11 monuments per 1000 km^2). Detailed data is provided in Fig. 2.

For the needs of a detailed inventory, based on data from the Report prepared by The National Heritage Board of Poland [10], archaeological monuments were divided according to its function. The following functional groups have been used:

- defensive (including strongholds, fortifications, etc.),
- ritual (including: churches, temples, chapels, monastic complexes, eremitical stone circles, places of worship, sacrificial sites, etc.),
- settlement / residential (including: manors, cities, palaces, architectural relics, settlements, villages, camps, caves, rock shelters, settlement traces, etc.),
- sepulchral-graves (including: burial mounds and megalithic graves),
- sepulchral-graveyards (including: burial grounds, megalithic burial grounds),
- industrial (including: infrastructure, places of raw material exploitation, production sites),
- various (including settlement complexes, mounds, shafts, multi-phase stations with different functions).

Detailed data on the number of individual types of archaeological monuments in the Warmian-Masurian Voivodeship is shown in Fig. 3.

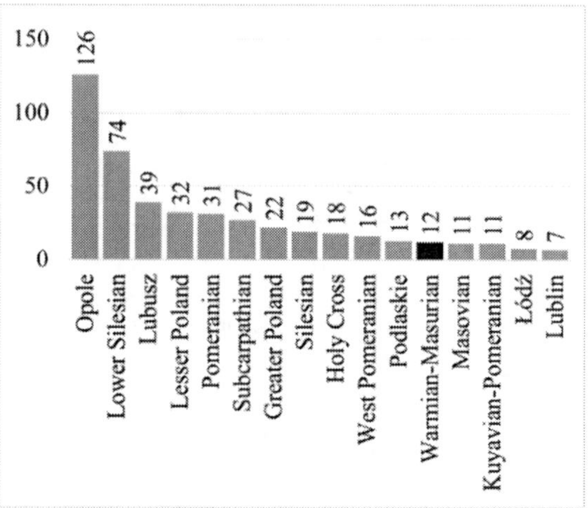

Fig. 2. The number of archaeological monuments per 1000km^2, divided into Polish voivodeships.

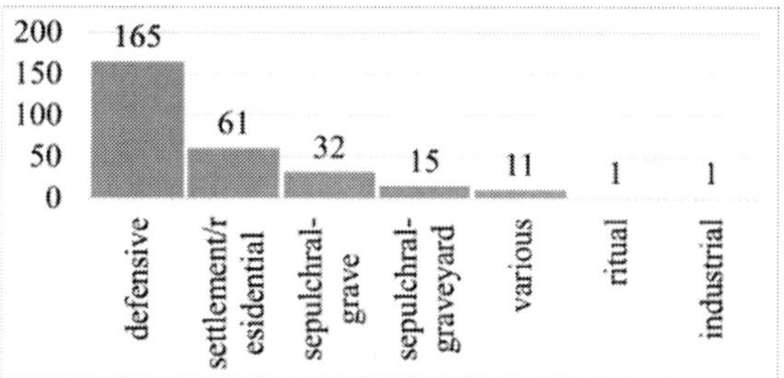

Fig. 3. Archaeological monuments in the Warmian-Masurian voivodeship, division by function.

Apart from the functional division, archaeological monuments can be divided according to the estimated time of their creation [10], for example: prehistory, stone age, bronze age, iron age, early middle ages, middle ages, modern era, etc. Detailed data on the number of archaeological monuments in individual chronological groups in the Warmian-Masurian Voivodeship can be found in Fig. 4.

National Heritage Board of Poland through geoportal (*mapy.zabytek.gov.pl/nid*) provides online information about objects entered in the register of objects of cultural heritage, free of charge. However, in order to carry out studies on the spatial distribution of the analysed elements, precise point data had to be obtained. Such data cannot be obtained automatically from the National Heritage Board of Poland geoportal, so it was necessary to have it manually collected.

Data on objects entered in the register of objects of cultural heritage as archaeological monuments were obtained from the Otwarte Dane portal (*dane.gov.pl*). In order to create a point layer necessary for analysis, the tabular data had to be associated with geographical coordinates. The original assumption was to analyse the entire country, but due to the amount of information needed (geographical coordinates of 7 672 points), as well as the universality of the methodology used, the area of the Warmian-Masurian voivodeship was analysed.

After supplementing tabular data with information on geographical coordinates, the database was completed with information on functional categories, previously defined. Information on the time of the objects' creation was also added. The prepared table was converted into a point cloud using the "Make XY Event Layer" of the ArcMap component of the ArcGIS software. After collecting quantitative and generic data, the next step was to analyse using GIS tools.

In order to determine the spatial distribution of the analysed objects, the Kernel Density function was used. Abandoning the limitations resulting from the territorial division allowed for a more detailed analysis [11-15], and the nature and age of archaeological monuments is exceptionally unrelated to the current administrative boundaries [16].

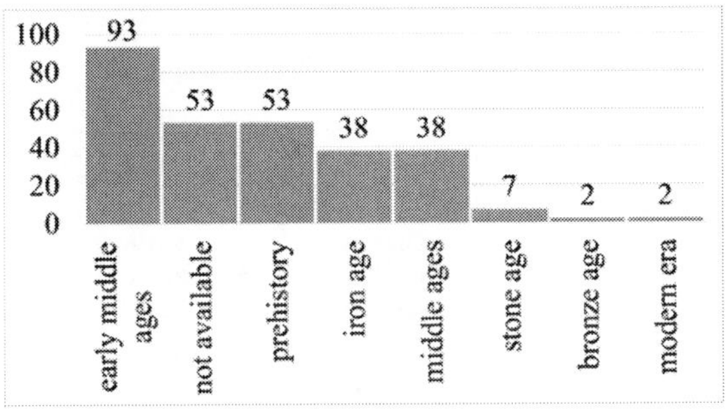

Fig. 4. Archaeological monuments in the Warmian-Masurian voivodeship, division by time of creation.

3 Results and discussion

The location of objects entered in the register of objects of cultural heritage as archaeological monuments, located in the Warmian-Masurian voivodeship, is presented in the form of a cartodiagram on Fig. 5.

The first stage of the study of saturation of archaeological monuments was to create a density map without division into functional or chronological groups, resulting in Fig. 6.

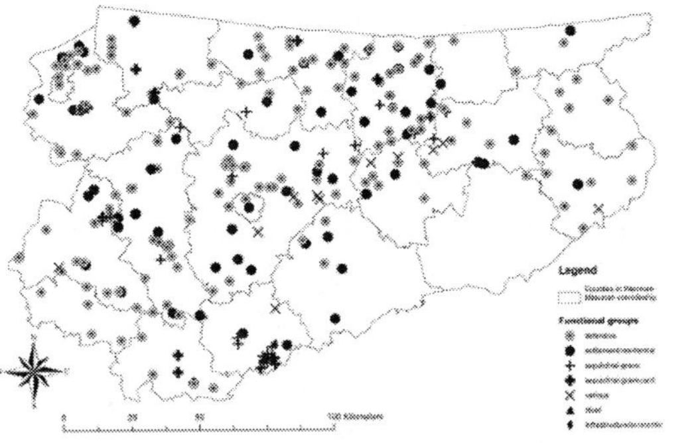

Fig. 5. Archaeological monuments, located in the Warmian-Masurian voivodeship.

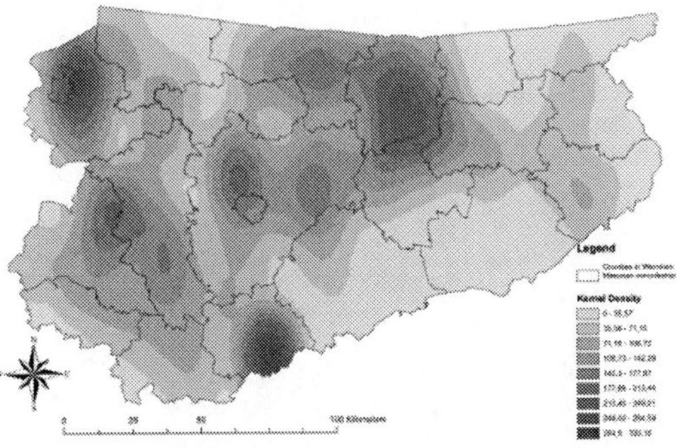

Fig. 6. Overall density of archaeological monuments in the Warmian-Masurian voivodeship.

The Warmian-Masurian voivodeship is characterized by a small number of archaeological monuments, however, they are fairly evenly distributed in the area of the voivodeship. The exception is the southeast of the voivodeship – there is no archaeological monument in the Pisz county.

The database, based on the occurrence of monuments in the area of the Warmian-Masurian Voivodeship, was divided into five functional groups. Four most numerous groups of objects were determinated: defensive, settlement/residential, sepulchral-graves and sepulchral-graveyards. Due to the small number of objects in groups: ritual and industrial (one object each), these were included in the group "various". The following figures successively shows the saturation of archaeological monuments of the functional groups: defensive (Fig. 7), settlement/residential (Fig. 8), sepulchral-graves (Fig. 9), sepulchral-graveyards (Fig. 10) and various (Fig. 11).

Because 115 out of 165 objects from the "defensive" function group are simultaneously in the chronological group "early middle ages" or "middle ages", the spatial distribution shown in Fig. 7 is not surprising – a large accumulation of objects near historic cities and fortresses can be seen (Giżycko, Kętrzyn, Olsztyn, Elbląg).

Due to the chronological diversity of objects from the "settlement/residential" function group, and lack of significant dominance of a specific historical period (7 objects were created in the stone age, 1 in the bronze age, 13 in the iron age, 1 in the early middle ages, 5 in the middle ages, 7 in prehistory and 1 in modern era),

 https://doi.org/10.1051/e3sconf/20186300019

and also due to the very general and basic nature of the group, Fig. 8 shows a similar spatial distribution as the general approach, excluding the southern part of the voivodeship.

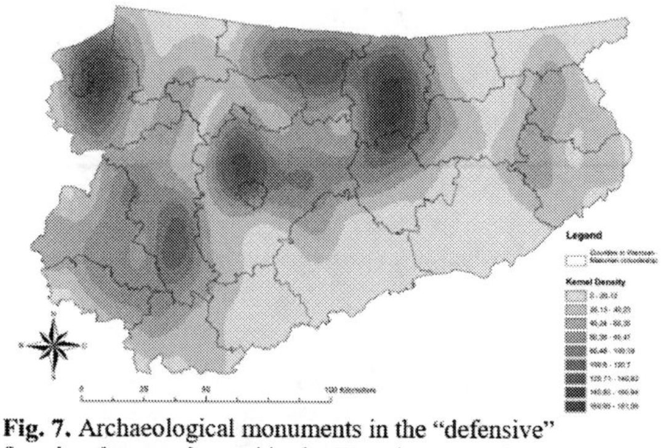

Fig. 7. Archaeological monuments in the "defensive" functional group, located in the Warmian-Masurian voivodeship.

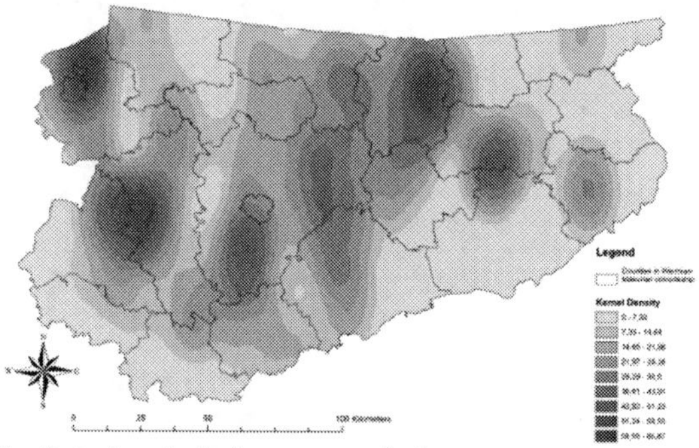

Fig. 8. Archaeological monuments in the "settlement/residential" functional group, located in the Warmian-Masurian voivodeship.

Fig. 9 shows the spatial distribution of the "sepulchral-graves" functional group. The Nidzica county is definitely dominating the distribution. 31 of 32 objects from this group were created in prehistoric times.

The spatial distribution of the "sepulchral-graveyards" function group is shown in Fig. 10. As in the case of Fig. 9, the largest density of objects from this group is located in the south of the voivodeship, in the Nidzica and Działdowo counties.

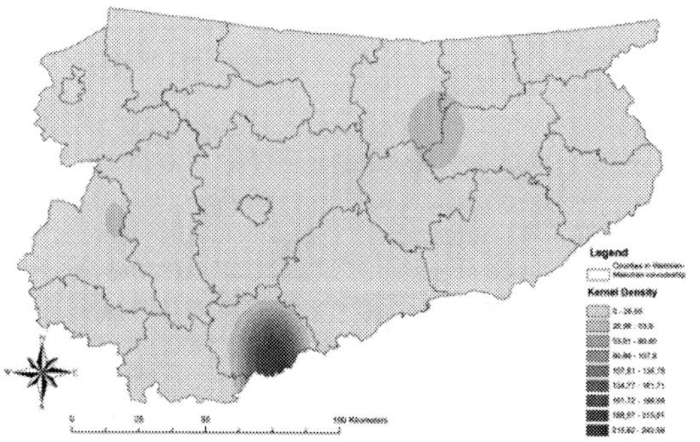

Fig. 9. Archaeological monuments in the "sepulchral-graves" functional group, located in the Warmian-Masurian voivodeship.

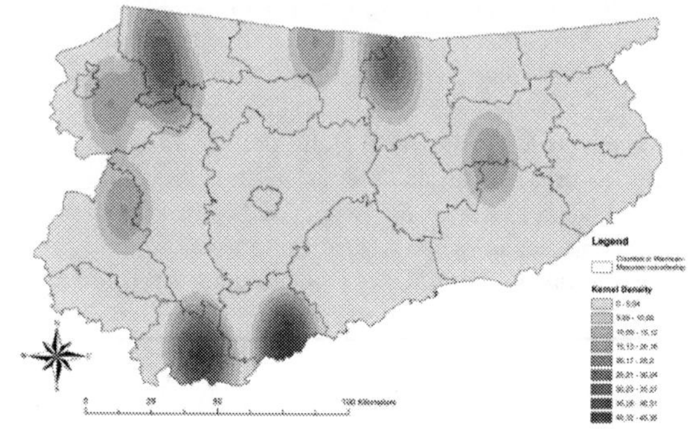

Fig. 10. Archaeological monuments in the "sepulchral-graveyards" functional group, located in the Warmian-Masurian voivodeship.

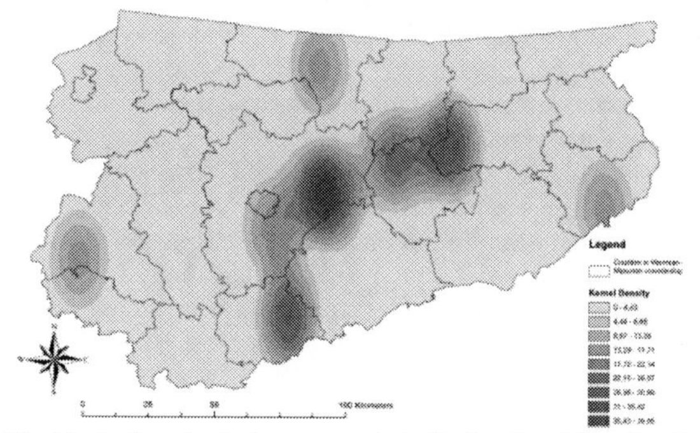

Fig. 11. Archaeological monuments in the "various" functional group, located in the Warmian-Masurian voivodeship.

Fig. 11 shows the spatial distribution of the "various" functional group. Objects from this group can be found primarily in the central part of the voivodeship.

In the final part, the periods of creation of the monuments were analysed, along with the location. In Fig. 12, using the cartodiagram, the chronological division of archaeological monuments in the area of Warmian-Masurian voivodeship is presented.

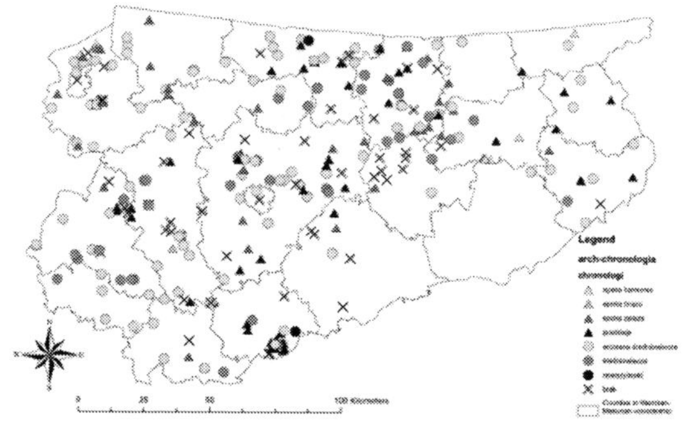

Fig. 12. Chronological division of archaeological monuments in the area of Warmian-Masurian voivodeship.

Due to the small number of objects in selected groups (7 objects from the stone age, 1 from the bronze age, 2 from modern era), Fig. 13, 14, and 15 used a generalized division into three groups: the oldest objects (from chronological groups: stone age, bronze age, iron age and prehistory), early medieval (from "early middle ages" chronological group) and the youngest objects (from chronological groups: middle ages and modern era). The analysis covered 233 objects, 53 archaeological monuments whose time of creation is impossible to determine, were omitted.

The largest density of the oldest objects can be seen in the southern part of the map, the area around today's Nidzica county. A spatial correlation can be seen with Fig. 9 and Fig. 10, depicting sepulchral objects.

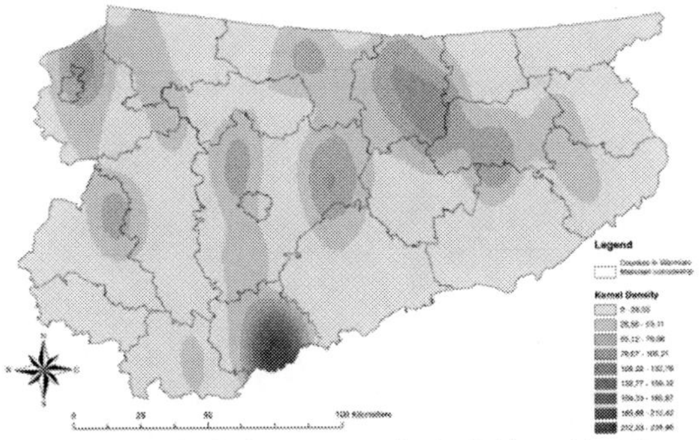

Fig. 13. Archaeological monuments in the "oldest objects" chronological group, located in the Warmian-Masurian voivodeship

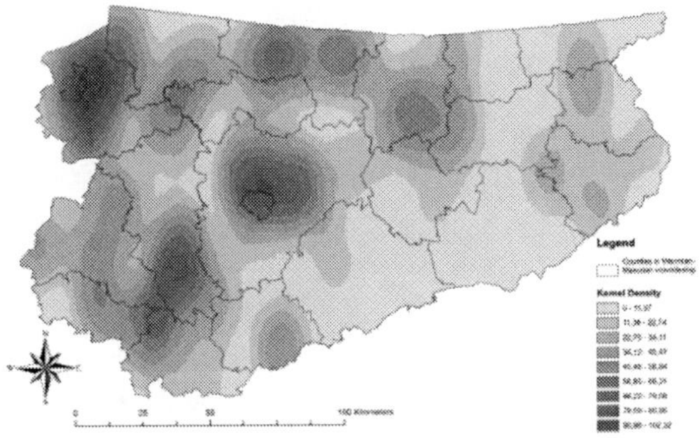

Fig. 14. Archaeological monuments in the "early medieval" chronological group, located in the Warmian-Masurian voivodeship

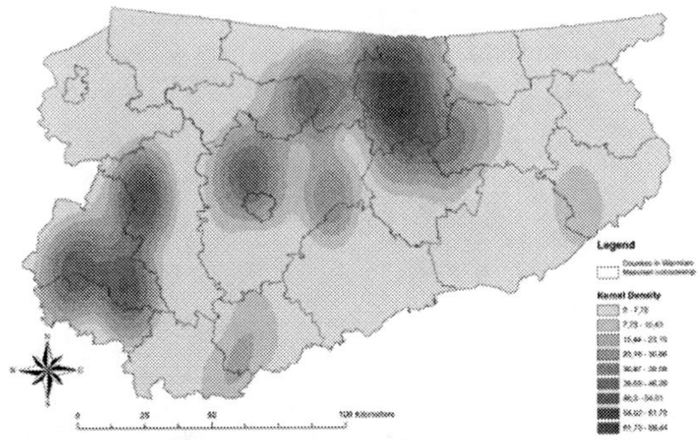

Fig. 15. Archaeological monuments in the "youngest objects" chronological group, located in the Warmian-Masurian voivodeship

Fig. 14 shows a spatial correlation with Fig. 6, showing the spatial distribution of all archaeological monuments in the Warmian-Masurian Voivodeship. This may be due to the fact that almost 42% of the analysed objects are from early middle ages.

Fig. 15 shows a spatial correlation with Fig. 7, which confirms the conclusions contained in the analysis under this figure.

4 Conclusions

1. Archaeological monuments, due to the age of its creation, indicate a complete lack of connection with current administrative borders. By abandoning the artificial boundaries and analysing the exact position of the examined objects, a more accurate view of the spatial distribution was obtained.
2. The conducted research allowed to conclude, that due to the need to limit the analysis in a unified and well-defined spatial framework, the boundaries of historical lands, not today's administrative boundaries, should be rather used. They should be determined on the basis of separate researches or analyses of historical maps.
3. The analyses made it possible to verify the spatial distribution of archaeological monuments in the Warmian-Masurian voivodeship, divided into various functional groups. Non-uniformity of their spatial distribution in the analysed area has been presented.
4. The analyses presented can be carried out both locally and regionally, for the whole country. They are universal in terms of area – can be made for any administratively, functionally or spatially separated area.
5. When creating the Municipality Development Strategy and Study of the Conditions and Directions of the Spatial Management of Municipality, local self-government must take into account the cultural and historical factors of the municipality. The presented analyses are a useful tool supporting the process of creating these documents. Similarly, it can be used in special cases when developing the local spatial management plans.
6. Local self-governments at the voivodeship level, in order to locate areas of exceptional historical value in a simple and quick way, may use similar analyses when creating and updating the Voivodeship Development Strategy and Voivodeship Spatial Management Plan.

References

1. The Act on the Protection of Monuments and the Care of Historical Monuments. Parliament. Journal of Laws from 2014, No 1446 as amended (2003)
2. P.M. Protopsaltis, *Book Review: Cultural Heritage in International Investment Law and Arbitration, by Valentina Vadi*. (Cambridge: Cambridge University Press). 35 Journal of International Arbitration, Kluwer Law International BV, The Netherlands. Issue **2/2018**, pp. 267–271 (2014)

3. E. Lagrange, S. Oeter, R. Uerpmann-Wittzack, eds. *Cultural Heritage and International Law: Objects, Means and Ends of International Protection*. Springer International Publishing, Basel. (2018) DOI:10.1007/978-3-319-78789-3
4. X. JIAO, J. WANG, *Context Protection and Inheritance in the Process of Urbanization*. 2nd International Conference on Education, Management and Applied Social Science (EMASS), pp. 170-174 (2018) DOI:10.12783/dtssehs/emass2018/20409
5. L.B. Kakashvili, *Archaeological museum-reservations as a cultural investment*. International scientific and practical conference world science ROST (Dubai). **2,** no 12 (28), pp. 49-51 (2017)
6. K. Rzasa, M. Ogryzek, M. Kulawiak, *Cultural Heritage in Spatial Planning*, Baltic Geodetic Congress (BGC Geomatics), Gdańsk, pp. 85-89, 2016, DOI:10.1109/BGC.Geomatics.2016 (2016)
7. K. Rząsa. Kulturowe aspekty planowania przestrzennego na obszarach wiejskich. *Kształtowanie przestrzeni wiejskiej*. UWM w Olsztynie. pp.139-154. (2017)
8. M. Ciski, K. Rząsa. Cultural Parks in The Spatial Planning System in Poland. *Baltic Geodetic Congress (BGC Geomatics)*, Olsztyn. pp. 145-149 (2018) DOI:10.1109/BGC-Geomatics.2018.00033
9. A. Momčilovic-Petronijević, P. Petronijević, M. Mitković. *Degradation of archeological sites – case study Caričin Grad. University of Niš, Serbia*. Facta Universitatis Series: Architecture and Civil Engineering **16**, No 2. (2018) DOI:10.2298/FUACE171201006M
10. Raport o stanie zachowania zabytków nieruchomych w Polsce, Narodowy Instytut Dziedzictwa, Warszawa (2017) (in Polish)
11. Z. Qin, W. Li, and X. Xiong, *Estimating wind speed probability distribution using kernel density method*, Electric Power Systems Research, **81**, no. 12, pp. 2139–2146 (2011)
12. E. García-Portugués, R. M. Crujeiras, and W. González-Manteiga, "Kernel density estimation for directional–linear data," Journal of Multivariate Analysis, vol. **121**, pp. 152–175. (2013)
13. J. Kim and C. Kim, *Reducing the mean squared error in kernel density estimation*, Journal of the Korean Statistical Society, **42**, no. 3, pp. 387–397 (2013)
14. M. S. Gerber, *Predicting crime using Twitter and kernel density estimation*, Decision Support Systems, **61**, pp. 115–125 (2014)
15. E. Saule, D. Panchananam, A. Hohl, W. Tang, and E. Delmelle, *Parallel Space-Time Kernel Density Estimation*, International Conference on Parallel Processing, pp. 483–492 (2017)
16. A. Fiedukowicz, A. Głażewski, A. Kołodziej, K. Koszewski, P. Kowalski, R. Olszewski, and L. Włochyński. *Map Portal as a Tool to Share Information on Cultural Heritage Illustrated by the National Heritage Board Geoportal*. Advances in Digital Cultural Heritage International Workshop Funchal, Madeira, Portugal, June 28. Springer International Publishing AG, Switzerland. pp.48-64 (2017) DOI:10.1007/978-3-319-75789-6

Low cost real time UAV stereo photogrammetry modelling technique – accuracy considerations

Pawel Burdziakowski *

Faculty of Civil and Environmental Engineering, Gdansk University of Technology, Narutowicza 11/12 , 80-233 Gdansk, Poland

Abstract. The paper presents accuracy considerations regarding three 3D modelling techniques. The tested new consumer type stereo camera (ZED 3D Stereolabs) has been implemented info an aerial mapping system, on board micro air vehicle MAV) and tested object has been mapped using a real-time photogrammetry with original real-time software application. The evaluated results has been compared with model gained with a state of art unmanned aerial vehicles (UAV) photogrammetry process using commercial UAV and commercial software, and the terrestrial photogrammetry modelling process with other commercial software. Papers concludes the tested real-time technology accuracy with compare to the traditional technique, and discusses real-time photogrammetry modelling in terms of engineering applications.

1 Introduction

A low cost consumer grade range 3D cameras market grew significantly last years. The evolution was a response on a recent demand from the gaming, entertainment and robot industry. The industry has boosted the development of low-cost consumer-grade cameras. The interest in these sensors initially remained at relatively low level, up to the moment when Microsoft company introduced Kinect sensor to their video-game console. The Kinect sensor is used as a motion input device. Over 10 million of Kinect units, flowed video game market, and has had a significant impact on the photogrammetric community. The adorability and versatility of the Microsoft's sensor, and similar devices, led to it's wide range of sensing applications such a 3D scene reconstruction, engineering tasks, robotic navigation, simultaneous localization and mapping techniques, and many similar applications, where time of data processing and acquisition is the main priority. Kinect sensor type (or other named device using the same PrimeSense chipset) have a relatively short (1-6 m) effective maximum working range, with limited accuracy (5-20 mm) that's significantly limiting a engineering measurements requirements. That kind of consumer grade cameras can be divided into two categories: range cameras based on triangulation and range cameras based on time-of-flight (TOF) principle. The Kinect technology can be described as structured light using a fixed pattern projection. The dot pattern is generated by a diffractive optical element and a near-infrared laser diode. TOF flight cameras use a single LED to emit sinusoidal modulated infrared light. A synchronized imagine chips takes four samples on one period of a sinus signal, what allows to compute the phase shift and consequently depth for each pixel [1]. The cameras described above are transmitting any kind of near-infrared pattern or signal into the object of measurement. The technology proved to be very prone to the strong daylight and not as effective in daylight conditions as indoor [2].

An indoor mapping and navigation become a challenge and highly desired technology for smart sites and robot applications [3]. The visual light sensors (RGB-D cameras) are playing a significant role within robotic community, closing any robots (humanoid robots, micro air vehicle (MAV), unmanned aerial vehicle) to humane like environment sensing [4].

Outdoor mapping, as mentioned above, is very challenging for a low cost RGB-D Kinect type camera. In the case of outdoor mapping and navigation a passive 3D cameras can be considered as a desired solution. For the presented research the ZED 3D Stereolabs camera has been chosen. The ZED 3D camera is based on passive stereovision, and consists of two RGB cameras with fixed base distance to 120 mm, which allows to generate depth image up to 20 meters (40 m is the maximum distance on the new updated firmware). The ZED 3D camera is optimized for real-time calculation using Nvidia CUDA technology. The ZED camera (among variety of options) can deliver already rectified stereo pairs in highest resolution (4416x1242) at 15 frames per second (fps). Lower resolution HD720 supports higher frame rate (60 fps) and data stream, what is recommended for visual navigation applications on fast moving robots (UAV, MAV) or mapping from the ground or surface moving platform. The ZED camera provides data via high speed USB 3.0 interface and is

* Corresponding author: pawel.burdziakowski@pg.edu.pl

https://doi.org/10.1051/e3sconf/20186300020

able to GPU's (graphical processing unit) accelerated real-time calculations. The camera outputs a real-time images in the form RGB rectified video (left and right), raw (unrectified) RGB video (left and right), left and right rectified side-by-side video (SBS), 32-bits depth map image, registered point cloud, confidence image and disparity image. The calculation is already done on the hardware level on the camera chip itself. The data can be recorded also on custom video format .svo and can be processed on the accompanying software.

The paper considers the ZED 3D camera accuracy mapping results, using Stereolabs application programming interface (API) and its standard camera calculations algorithms in terms of engineering 3D measurements and modelling, compared to the typical UAV mapping technique and terrestrial photogrammetry with digital single-lens reflex camera (DSLR). All three techniques have been used discreetly for the modelling of the same 3D object in outdoor environment in daylight conditions.

Fig. 1 ZED Stereolabs 3D camera.

2 Data acquisition and processing

The object has been placed on the outdoor environment, marked with 12 bits coded targets as a control points for scale recovery and dimensions control. The first modelling technique utilizes MAV (DJI Mavic Pro) with Bentley Context Capture Software for data processing (georeferenced UAV mapping). The MAV (Fig. 1a). At the moment, this product is the smallest flying and stabilized camera on the commercial market (weight only 734 g). The platform utilizes for navigation one GPS and GLONASS module, two Inertial Measurement Unit (IMU) modules, Forward and Downward Vision System to automatically stabilize itself and navigate between obstacles and track moving objects. The presented MAV is using non-metric cameras with the sensor size 1/2.3`` - 6.17x4.55 mm with the pixel size 1.55 μm. The camera of this MAV tags (into the EXIF metadata) the images witch geolocation data using MAV's GPS (direct image georeferencing). The modelling software, Context Capture, is the reality modelling category application for processing images of the physical environment into 3D representations to provide current context within geospatial modelling environments.

a) *b)* *c)*

Fig. 2. a) DJI Mavic Pro, b) Canon EOS 500D, c) AI-GEO MAV with ZED Camera.

The second modelling technique utilises DSLR Canon EOS 500 D camera with standard kit lens (EF-S 18-55mm f/3.5-5.6 IS) (Fig. 2b) (terrestrial photogrammetry). Cannon camera is equipped with CMOS sensor with effective 15.1 million pixels. For this task, the Agisoft PhotoScan Pro has been used, as a data processing software. Agisoft is a stand-alone software product that performs photogrammetric processing of digital images and generates 3D spatial data.

The tested configuration, ZED Stereolabs camera, was implemented on board original developer platform named AI-GEO MAV [5] (Fig. 2c). The platform is designed for real-time photogrammetry applications and employs NVidia TX2 embedded commuter for real-time photogrammetry calculations. The operating system Linux Ubuntu running the original software based on Stereolab's application programming interface (API). In this case API is a set of subroutine definitions, communication protocols and tools for building original software, or generally it is a set of clearly defined methods of communication among various components within the system. API's algorithm have not been modified for this research and used as delivered by the Stereolab's within own original mapping commuter application. It has to be mentioned, that the model has been calculated based on the Stereolab's delivered API without any API's code modification. The code is dedicated to the real-time data processing, so it can be expected that accuracy of the model accuracy part can be lower. The presented approach allows evaluate the quality of the overall real-time mapping process. Oppositely to the research [6] where authors evaluated the same camera but using the photogrammetry software, the quality of the models was comparable to the typical UAV photogrammetry technology. Here, in this research, the camera and its API are evaluated in real-time UAV mapping application.

2.1. Georeferenced UAV mapping

The first model calculated using low altitude photogrammetry methods. The images were taken around the object (Fig. 3a). The MAV size allows to carried out the flight around the object in the presented localization.

a) b)

Fig. 3. UAV photogrametry, a) flight patch and extrincis image parameters, b) point cloud model.

The object was modelled (Fig. 3b) form 51 georeferenced images (4000x3000) using Bentley Context Capture. The software reported reprojection error (RMS) is 0.71 pixels, with a minimum resolution

 https://doi.org/10.1051/e3sconf/20186300020

of 0.034 mm/pixel and a maximum of 1.564 meters/pixel. The median resolution equals 0.191 meters/pixel. The model has no shortages, all surfaces has its representation in point cloud. Due to above reason the model form UAV photogrammetry, for this research, has been considered as the referenced model.

2.2 Terrestrial photogrammetry

The object was situated in the place where the terrestrial photogrammetry technique was inconvenient to use. The outer part of the object was unable to photograph without man lift, nevertheless the images offered the highest resolution, in the presented study, the problems with a camera position caused acquisition problems and some part of the object has not been photographed. The 12 bit circular coded control point has been established and measured. The scale of the object was recovered using marked tie points measurements. The same points was used to the accuracy control in the UAV photogrammetry case.

a) b)

Fig. 4. Terestrial photogrametry, a) camera position and images orientation, b) point cloud model.

The object modelled using terrestrial photogrammetry (Fig. 4b) form 74 images (5184 x 3456) using Agisoft Photo Scan has some shortages, and surfaces has no representation in point cloud. The top of the modelled object has not covered by the source images, due to the object location. All recovered surfaces has a quality and accuracy adequate to the referenced model. Software reported reprojection error (RMS) is 1,99 pixels.

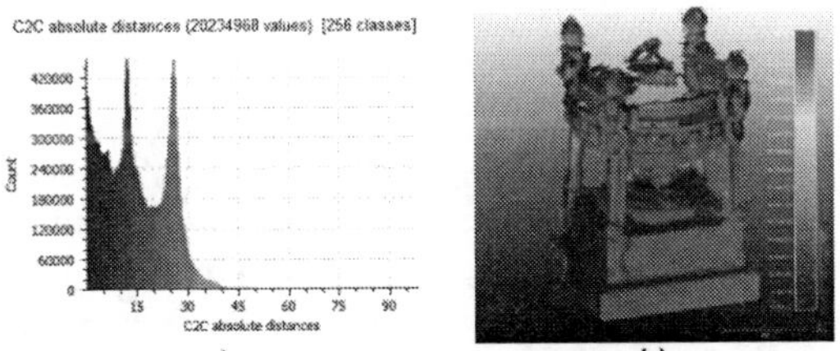

a) b)

Fig. 5. Cloud to cloud distances comparison (DJI to Canon) a) histogram, b) coloured model and scaled distances in mm.

The comparison of the referenced cloud form DJI Mavic Pro (UAV photogrammetry) to the cloud form Canon camera (terrestrial photogrammetry) (Fig. 5), revealed that the max distances not exceed approx. 35 mm. It confirms the initial assumptions, that the UAV model can be considered as a reference model.

2.3 ZED 3D modelling

The real-time UAV mapping technology has been applied on the original developer UAV platform [5]. The platform allows to deliver the mapping instrument (ZED 3D Camera) to the desired area or object with simultaneous data acquisition and processing on board the platforms commuter. The operator receives processed data in point cloud or meshes (textured, only surface or triangle mesh). The Fig. 6 presents the main interface of the real-time mapping software. The visual light camera image is presented to operator's overall orientation in the vicinity of the object of interest. The recovered depth image in the 12 bit colour scale allows to monitor the quality of the stereo matching process and allows to adjust the camera parameters (the mapping distance). The live meshing process shows to operator already mapped environment and acquired data. In case the environment has not been mapped, there is a possibility to repeat the flight in the not mapped areas.

Fig. 6. The real-time mapping software live view (left top: camera view, left bottom: disparity map (depth image), right: life meshing process).

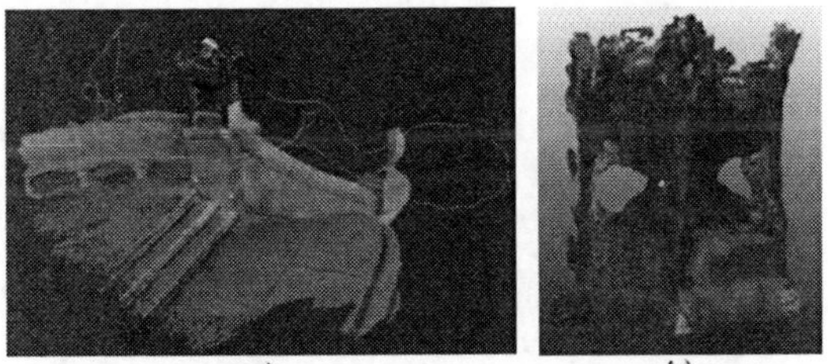

a) b)

Fig. 7. The platform visual odometry flight patch (a) and textured model of the object and form ZED Stereolabs (b).

The object modelled form the ZED camera using MAV presented on the Fig. 7b. The data has been acquired as a side by side video with resolution HD720 and 60 frames per second. The left and right video frames are synchronized and streamed as a single uncompressed video frame in side-by-side format (resolution for SBS video 2560x720) to the on-board processing commuter.

The texturing process time not exceed 2 minutes, for the mapping area, and starts after the data gathering is finished and the operator confirms the area has been mapped. The point cloud and mesh is received in the real time, means that are created during the flight and results are displayed on the operators screen, but the texturing process stars always after the complete mesh is

 https://doi.org/10.1051/e3sconf/20186300020

recovered. A texture map is generated for the whole area.

The models point to point comparison clarifies the main modelling results. The tested equipment result has been compared to the reference model (Fig. 8).

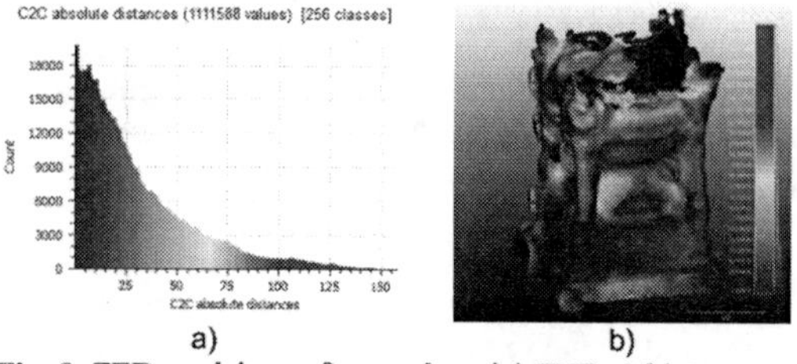

a) b)

Fig. 8. ZED model vs referenced model (DJI) a) histogram, b)coloured model and scaled distances.

The comparison of the referenced cloud to the cloud form real-time modelling process using ZED Camera (Fig. 8), revealed that the max distances not exceed approx. 125 mm. The biggest deviation from the referenced model can be observed in the areas where the camera did not acquire a sufficient data (above 100 mm), covered by other object elements (approx. 60-75 mm), very detailed object part (approx. 50 mm). Deviations below 25 mm can be observed on the flat and simple shape object parts. It can be noticed, that the object overall shape and middle size details are modelled. The smaller details, are simplified on the mesh and clearly not modelled (50 mm deviation). The mesh generation process simplifies the details, and fast data processing are at the expense of details quality.

4 Conclusions

A real-time photogrammetry applications and technique are still under the development. A fast computation techniques allows to run calculation process, even that complicated, in the real-time on board micro aerial vehicles. The paper considered three different modelling techniques, form high resolution terrestrial photogrammetry, thorough offline commercial UAV photogrammetry, and finally to the real-time stereo photogrammetry form the unmanned aerial vehicles. It can be stated that the presented solution can deliver the spatial data in the real time, means during the flight. The data can be analysed and monitored online, via wireless connection. The spatial data are delivered already in the popular interchangeable file format (*.obj, *.las), and can be used in third party software on in own applications. Form the photogrammetric point of view the quality of calculated the model is affected by the fast data calculation priority. Consequently, this research opens the discussion, if the fast 3D camera volume measurements are meet the geodetic standards. As it can be showed here, the very detailed geometric modelling is rather impossible using the API delivered by Sterolabs at the geodetic grade level. Despite the fact that presented API simplifies the object, and detailed are lost, it would be possible to use either a different software for real-time spatial data calculation for a very detailed object, or use this process for the larger object i.e. for geodetic volume calculation.

The visual textured model quality, for the user, presents a very attractive form (Fig. 7a). The quality of this process mitigates the geometry weakness. Texturization exposes all small details as a picture on the object, and consequently the user can have an immersion that the details have been modelled and have their geometrical representation in point cloud or mesh.

The presented mapping results have a quality and accuracy on the navigation level, means that the model of the environment can be safely used for the autonomous UAV navigation applications for example on board unmanned platform [7], and for initial and rough geodetic mapping (using standard camera API). While the photogrammetry offline software is engaged, the quality of the results would be on the typical photogrammetric accuracy level, however calculation time will be affected. The good solution would be also to use online photogrammetry technique with live data calculation on remote server and stereo image transmission via internet connection [8] and methods like [9].

The significant problems has been observed while the UAV moved to the shadowed area and the cameras parameters has not been automatically (or wrongly) adjusted. In that case the radiometric correction [10] should be implemented in the real-time data acquisition process. The moving objects with the sensing equipment (cameras) are not always faced the same light condition on the whole object, as well as on the background. That was observed in this case. The online, real-time radiometric correction should be implemented.

Some not modelled parts of the object lead to the conclusion, that the construction of the AI-GEO MAV [5] and 3D camera mount should be modified. The tilting mechanism should be added for mapping an object underneath the MAV while the platform is loitering in the one position or circles around.

References

1. T. Luhmann, S. Robson, S. Kyle, and J. Boehm, *Close-range photogrammetry and 3D imaging*. Walter de Gruyter (2014)
2. F. Endres, J. Hess, J. Sturm, D. Cremers, and W. Burgard, *3-D Mapping With an RGB-D Camera*, IEEE Trans. Robot., **30**, no. 1, pp. 177–187 (2014)
3. P. Burdziakowski, A. Janowski, M. Przyborski, and J. Szulwic, *A modern approach to an unmanned vehicle navigation*, 16th International Multidisciplinary Scientific GeoConference SGEM 2016, Book 2, **2**, pp. 747-758 (2016) ISBN 978-619-7105-59-9 / ISSN 1314-2704
4. T. Gupta and H. Li, *Indoor mapping for smart cities - An affordable approach: Using Kinect Sensor and ZED stereo camera*, 2017 International Conference on Indoor Positioning and Indoor Navigation (IPIN), pp. 1–8 (2017)

5. P. Burdziakowski, UAV design and construction for real time photogrammetry and visual navigation, 2018 Baltic Geodetic Congress (BGC Geomatics) (2018)
6. A. Deris, I. Trigonis, A. Aravanis, and E. K. Stathopoulou, *Depth cameras on UAVs: A first approach*, International Archives of the Photogrammetry, Remote Sensing and Spatial Information Sciences - ISPRS Archives, **42**, no. 2W3, pp. 231–236 (2017)
7. A. Stateczny, D. Gronska, and W. Motyl, *Hydrodron-New Step for Professional Hydrography for Restricted Waters*, 2018 Baltic Geodetic Congress (BGC Geomatics), pp. 226–230 (2018)
8. Z. Paszotta, J. Szulwic, and M. Szumilo, *Internet Photogrammetry for Inspection of Seaports*, Polish Marit. Res., **24**, no. s1(93), pp. 174–181 (2017)
9. M. Kedzierski and P. Delis, *Fast orientation of video images of buildings acquired from a UAV without stabilization*, Sensors (Switzerland), **16**, no. 7 (2016)
10. M. Kedzierski and D. Wierzbicki, *Methodology of improvement of radiometric quality of images acquired from low altitudes*, Meas. J. Int. Meas. Confed., **92**, pp. 70–78 (2016)

EDP Sciences – Web of Conferences
17, Avenue du Hoggar
Parc d'Activité de Courtabouf
BP 112
F-91944 Les Ulis Cedex A
France

ISBN 978-1-5108-7477-0

CURRAN ASSOCIATES INC.
proceedings.com

Twelfth International Conference on Advances in Computer-Human Interactions (ACHI 2019)

Held at DigitalWorld 2019

Athens, Greece
24 – 28 February 2019